CRISIS

CULL or

COUP?

WHAT, HOW & WHO?

FACTS & TRUTHS TO MAKE YOU THINK

Exposing the Great Lie
& the Truth about the Covid-19 phenomenon

CRISIS, CULL OR COUP: WHAT, HOW & WHO?
Facts and Truths to Make You Think

ISBN: 978-1-906628-772

Published by CheckPoint Press, Ireland.

www.checkpointpress.com

About the Author: At the time of writing, Stephen T Manning is the Administrator of the *Integrity Ireland Association,* and an Officer of *The Peoples Tribunal of Ireland.* Both bodies were established (in 2012 and 2020 respectively) to challenge endemic corruption, cronyism and criminal cover-ups in agencies of the Irish State, especially within the Irish justice system. In pursuit of this voluntary work, Stephen has gained an intimate insight into the workings of 'the establishment'.

Prior to this, Stephen taught at a number of schools, colleges and Universities in Europe, Asia and the United States, and previously worked as a mountain sports instructor and personal trainer following NATO military service. He holds qualifications in a number of fields, most notably in sport, psychology and the study of world religions, and maintains an active interest in truth-and-justice-related matters that affect us all. He is the author of a number of books including:

- *Psychology, Symbolism & the Sacred; Confronting Religious Dysfunction in a Changing World.*
- *The Colour of Truth Vol I: Amazing Coincidence? ...or Intelligent Design? (Full colour version ISBN: 978-1-906628-376)*
- *The Integrity Ireland S.O.S. Guide (Saving Our State).*
- *D.I.Y. Justice in Ireland. Prosecuting by Common Informer.*
- *The Peoples Tribunal of Ireland, Handbook, V.1.*
- *Criminality in the Irish Courts & the Absence of the Rule of Law.*
- *INDICTMENT – & Petition for a Public Enquiry into State-sponsored Criminality in Ireland.*

Soon to be published:

- *"The Rise of the Sociopath and why YOU should be worried!"*
- *"Lifting Mary's Veil: Unknown Secrets of the Feminine Divine."*
- *"The Colour of Truth" (Volumes II & III)*
- *"JUDGE-HUNTING: & Other Fun Ways to Protect Your Rights."*

For further information, please visit www.checkpoint.ie

CONTENTS

TO THE READER

This booklet has been composed for the purposes of raising an informed, fact-based and objective review of "The Covid-19 Phenomenon" that incorporates 'the official narrative' as well as certain qualified opposing positions and theories; to be jointly explored in context as to their credibility or factual accuracy.

It is hoped that by raising a well-informed but hopefully easily-understood and logical discussion, that the undeniable truth will emerge, thereby alleviating much of the angst, confusion and contentious debates that are polarising societies and dividing families, as the very real effects of 'the Covid-19 phenomenon' bite into each of our experiences of daily life.

Thousands are dying because of the Covid-19 phenomenon. Many millions more are facing into a future where life hardly seems worth living. The longstanding truism that, "Truth is its own witness" may never have had such import, and we owe it to ourselves and to future generations to have this debate now; and to have it in a well-informed context, honestly, intelligently and courageously.

For us humans, 'being alive' means more than just functioning as a biological organism, intelligent or not as the case may be; it means being truly alive and aware as to the purpose and meaning of life.

Our sincere and enduring hope is that this book will help to inform, and will help to save lives – in more ways than one might imagine.

Online links have been included as endnotes for the purposes of informing the reader of the collective public debate, and as references to some of the factual contents. Any such quotes or contents, including any inaccuracies, bias, errors or offensive opinions in those linked materials should be understood in context of the respective authors' stated positions.

INTRODUCTION

What are the origins and effects of the Covid-19 phenomenon? What exactly is it? Is it a real disease – or something else? What or who caused it, or created it? What precisely is the danger here to individuals and to society? Is it really as serious as they say? Why must we all wear facemasks, suffer lockdowns and get vaccinated? What's the difference between a normal coronavirus and SARS-CoV-2 and what is Covid-19's natural, predicted and/or pre-planned trajectory? What do viruses do anyway, and how do we usually deal with them? Is this a genuine global pandemic – an unexpected and unanticipated health crisis, or, some contrived-and-orchestrated *plan*-demic, and if so, who is driving the narrative and pushing the plan? And what of the increasing numbers of anti-vaxxers and so-called conspiracy theorists, the outspoken questioners and cynics? Why are they being censored and silenced? Should we even be listening to them? Are the PCR tests reliable or not? Do the new vaccines work? Now that we have more facts, what does the science and the data really say? Is Covid-19 now on the way out, or is it getting worse? What are these new 'variants' for example, and how do they fit into the story – and who is saying what? Is this all about the science and the physical facts, or is there some 'spiritual' or providential aspect to what is going on, or, is that just too fantastic a notion to consider? What then of the psychology and the emotional and mental damage being caused? How and why are some aspects of the economy thriving, whilst others are being bankrupted? Who are the big losers and winners? Is the virus that intelligent and discerning – really? Why all the trauma and enforced restrictions – and what of all the official contradictions? Was all this social and economic disruption really necessary? Who exactly are we getting our information from – and can they be trusted? Is the establishment really looking out for us, or is something else going on?

In short; what does this whole "Covid-19 phenomenon" actually comprise of? What are its various parts – and the sum of those parts – and what's *really* happening to our world? Where are we going with all of this upheaval – and is it really necessary? And if so, necessary for whom and for what? How does this affect me as an individual? Do we have any say in what's happening – and if so, what's my personal role and responsibility? Do we simply accept and comply – or should we be asking more questions? What's happening to our cultures, our societies and our families? How best to understand it all; how to deal with the repercussions and the effects that any of these variables could have on the economy, on human society and on future generations, and most especially on our fundamental rights to liberty, to education; to freedom of movement, of choice, and expression; and, to a 'normal' life of personal fulfilment and happiness?

The plain fact of the matter is that most of us have been broadsided by the sheer scale of the Covid-19 phenomenon, and many of us are still reeling with shock, fear and confusion as we struggle to get our heads around it all. The easy option of course is to accept the official narrative without question and meekly comply with all of the directions and diktats coming down the pipeline, especially when punishments and sanctions are already in place if we do not.

The contrary position is that something truly awful and terrible is happening to us, and it's not 'the virus' that we should be most afraid of, but rather, the sinister emergence of totalitarian measures that could spell the end of free-and-democratic society as we knew it just eighteen months ago. The move to radically change society—however it's being justified—from one where individual rights are honoured and respected, to a society where arguably, we will have none of those rights or privileges, must be resisted and questioned with every fibre of our being if we are to retain our essential humanity. And "being human" means more than just obeying orders, or earning a living, or managing to survive for 70 or 80 years in a world devoid of hope.

As with any great upheaval or catastrophe in history, it is our human response to it that shapes and defines the future that follows. That future then becomes our present, and then our collective past and so on, and thus (the old saying goes) we learn the lessons of history. Or do we? Because this is where another maxim comes into play, that, *"Those who fail to learn the lessons of history are condemned to repeat them!"* The emphasis has always been on learning, and understanding, and on properly educating ourselves so that we do NOT repeat the sins of our forefathers so-to-speak. But true education is not just about the mere collection of mathematical facts and the rote memorisation of historical data, it is also about understanding the context within which that data emerged and how it is being presented to us, and about acknowledging our own subjective biases, prejudices and pre-formed opinions as we go doggedly searching for 'the truth'.

In a world of vastly different personal opinions, collective ideologies and political perspectives however, 'the truth' can prove an elusive concept to nail down, especially in our new technological societies where information flows so fast. It was hard enough to get a rounded education in the old days, and difficult enough then to discern the truth in the mainstream media, but now, we also have to sift between false facts and fake news; through corporate advertising and government propaganda; and through conspiracy theories and social media outlets which each have their own take on 'truth'. This is why discernment in times like these is so hugely important and indeed absolutely vital to any proper understanding of what is going on. But it must be a well-informed, intelligent and courageous discernment that is rooted in truth itself, because that will be the only way that we can eliminate the lies.

In my own experience, there are two main ways to determine what is essentially true, and what is not. The first method is to be told a truth which you either accept on face value (which is never a good idea by the way) or, you put it to the test – somehow – which is only possible of course if you have the time, knowledge and resources to do so. Unfortunately, in the case of Covid-19 many of the 'truths' being offered to us by the medical-pharmaceutical industry, the mainstream media and our governments for example, are such that they require unquestioned acceptance and an immediate response (masks, tests, lockdowns) which, if those 'truths' later prove to be false, means that we are all participating in a lie – and that too is never a good idea.

The other way to establish the truth of any given matter is to simply identify all that cannot possibly be true, and eliminate it from consideration. The truth (or the best part of it at least) thus emerges.

This book has been written for anyone and everyone who wants to "see the bigger picture" here, especially in context of a broader understanding of what's right and what's wrong; of truth versus deception; and of good versus evil. It aims to inform the reader of the fundamentals and facts of the Covid-19 phenomenon against a backdrop of humane principles in added context of each person's responsibilities as a member of society. By the time you have finished reading we hope that the Covid-19 landscape will be clearer to you, and that you will have a better understanding of what is really going on. From there, you must evaluate what best to do – with the emphasis on 'best' (as opposed to easiest). One of the most immediate and pressing things that all moral people should be doing in the midst of this 'novel crisis' is to encourage each other to ask questions, and, when the answers just don't add up, to listen to our own intuition – to apply our God-given gifts of discernment and wisdom – and then simply, but courageously, to "do the right thing".

To those of my personal acquaintance who have, until now, held different or variant opinions sufficient to have caused any tensions or estrangement between us; I offer you this book, a token of affection and concern; as an olive branch extended with love and forgiveness (as always) and ask simply that you allow truth to be its own witness.

A word of caution to those of you who live by the mantra; *"I don't want to know!"* My respectful advice to you is NOT to read this book.

PART ONE

FOUNDATIONS

CHAPTER ONE

THE BASIC FACTS

ABOUT THE COVID-19 PHENOMENON

At the grave risk of starting this book by insulting everyone's intelligence, but at the same time being driven by the editorial impulse to lay out the argument clearly and succinctly from the start, let us clarify that we are using the somewhat ambiguous term 'the Covid-19 phenomenon' as opposed to some of the more specific terms available, so as to underline the essential fact that many of the supposed 'facts' that have bombarded the public airways, TV screens and news outlets since the emergence of the virus (in the mainstream media at least) are not in fact, 'facts' at all.

> **"FACT"** *(noun)*
> *(i) A thing that is known or proved to be true.*
> *(ii) Information used as evidence or as part of a report or news article.*
> *(iii) The truth about events as opposed to interpretation.*

Having opened with this no-doubt disconcerting observation (for some) the astute reader will already have guessed the general direction we are heading in; and 'NO' (for the possibly not-so-astute) this book is NOT going to be a wholesale endorsement of the official Covid-19 narrative. Far from it. This book is all about *genuine* facts and about provable truths, and about logical and rational conclusions that arise from an informed understanding of those facts-and-truths in context of 'the bigger picture' – with our blinkers removed. The course and direction of this Covid-19 phenomenon – and the end result (whatever that will be) depends largely on us knowing what's really going on. So, let us be clear and emphatic from the very start; that anything and everything that is presented as 'fact' in this book, will absolutely align with those three dictionary definitions above. Given the size and breadth of the topic however, and because of the great abundance of conflicting data and differing opinions already out there, we'll begin the discussion by sticking strictly to those facts that we are *all* in agreement with before looking more closely at the more contentious how's, what's, why's, where's, when's and who's of each major development, and in doing so, we should be able to eliminate much that is patently untrue or just not credible, whilst examining what's left so as to come up with *the* most likely explanation for what's really going on – and from there, to consider our options.

So, what *are* the basic historical facts? Well, the first fact to be established is that the world was first (officially) informed of the arrival of a new, highly contagious and potentially deadly virus on December 31st 2019. We all know the story by now. Subsequent reports from the Chinese authorities would claim that the first case of infection from this 'novel' (new) SARS-like coronavirus occurred in mid-November 2019 in Wuhan, China, but for reasons that still haven't been properly explained, the Chinese authorities (it seems) chose NOT to alert the world at that time (for about 6 weeks) despite believing (we are told) that the world was about to be hit by a deadly global pandemic. We will look closer at that scenario in a moment.

The next incontestable fact is that the world was informed, quite specifically by the Chinese authorities on January 7th 2020, that the virus would be *temporarily* named "2019-nCoV" indicating, quite simply, that they had identified a new ('n') corona-type virus ('CoV'). This was followed up by a more specific definition on February 11th 2020 when *The International Committee on Taxonomy of Viruses* (ICTV) renamed it, "SARS-CoV-2" meaning 'Severe Acute Respiratory Syndrome Coronavirus No. 2.' This indicated that this new virus was closely related to "SARS-CoV" (version No. 1) that was responsible for the relatively-modest SARS outbreak of 2003. Whilst most of us remained boggled by the science at the time, the world was now reassured that at least the scientists knew what we were dealing with. Meanwhile, as panic spread across the airwaves, diverse claims, accusations and theories about how the virus came about and who was responsible abounded. Was it a naturally-occurring event that spread from bats and pangolins to humans, or, was it deliberately manufactured in some Frankenstein-type laboratory? Were the Chinese ultimately responsible for the virus release (accidental or otherwise) or, should we believe official Chinese claims that it was the American Military who were to blame, and that this was an insidious, pre-planned geo-economic-political attempt to discredit China? Or, was this event far more sinister than that? Was this some sort of bio-weapon or a psychological tool of mass-compliance, deployed by 'the establishment' and by technocratic elites as necessary to 'The Great Reset' and other Machiavellian plans to create a New World Order?

The unfortunate reality is that we were never likely to get solid, empirical answers to the questions of the virus' true origins, because so many 'big players' were involved at the beginning laying blame at each other's doorsteps, and we all know how very hard it is for the establishment to back down or admit they were ever wrong. Even now, over eighteen months after the crisis began, the same big players – and now many more besides – all have a vested interest in controlling a narrative where so much is at stake, politically and economically, especially in terms of public trust; not to mention public compliance with the official narrative and/or public accountability for the drivers of that official narrative should some of the worst-case scenarios prove to be true. Even as this book was being composed, data that was never supposed to have become public has started to leak out of an otherwise very carefully managed script, and this in turn informs us that even though we may never get a definitive, comprehensive, truthful answer from the authorities as to how the crisis 'originated', that we shouldn't stop asking important questions either, because there are a number of highly-provocative clues and intriguing circumstances surrounding all of this Covid-19 phenomenon that need to be considered. Those key issues, we'll return to shortly.

The final fact that we need to establish is somewhat stating the obvious; that following the various official declarations that we had 'a highly-contagious pandemic' on our hands, that the world collapsed into a frenzy of panicked reactions – mostly by the various authorities, both medical and political – to ostensibly prevent what was being labelled as a prospective human catastrophe of almost-unimaginable proportions. The virus (we were told) would take the lives of many, many millions, if we did not respond immediately with draconian measures such as enforced lockdowns, facemasks, social distancing and small business closures; whilst submitting ourselves to tests, intrusive tracking-and-tracing procedures, and rushed injection programmes with what appears to be, 'not-properly-tested' vaccines.

As a direct result of these measures, hundreds of thousands of small-and-medium businesses have since collapsed or are on the verge of bankruptcy; public services (apart from Covid-related 'initiatives') have all but shut down; and countries all over the world have been plunged into insane levels of debt stretching way into the future. Big business on the other hand, and especially multinational corporations appear to be thriving. Free movement and casual socialising is all but banned, with the long-predicted deployment of 'Covid passports' now entering the game. Societies everywhere have entered a zombie-like state of existence that is undermining not only the physical, psychological and indeed spiritual health of individuals, but also of families and communities. Fundamental rights that we had previously all-but taken for granted, such as the right to freedom of movement, of travel, of speech or to earn a living, have been removed or suppressed with alarming efficiency, leaving most people variously confused, fearful and indeed increasingly angry at being dictated to by petty officials and autocratic governments – many in the pockets of big business – who dish out contradictory 'facts' coupled with asinine diktats, orders and edicts, laden with foreboding consequences, with infuriating and patronising arrogance. Everyone except the protected elites it seems, has had their lives violently interrupted, and, unless you are one of those people who believes that masks and injectibles and even more restrictions are 'the solution' then it's increasingly difficult to see where all of this ends.

Amongst the latest disturbing developments in mid-2021 are the introduction of social segregation, and the banning of non-vaccinated persons from international travel or indoor dining for example, as well as publicly-declared threats by certain first-world governments (USA, Australia and New Zealand for instance) that they may have to, "go door-to-door looking for unvaccinated persons" so as to "protect the general population". And all of this in face of overwhelming evidence of serious, technical, medical and ethical problems with the global Covid-19 vaccination program, including thousands of consequent deaths and hundreds of thousands (if not more) serious adverse reactions. For even more disturbing reasons that will soon become clear, this obviously-crucial information is largely being censored and suppressed by the political establishment and the mainstream media.

And that basically, summarises the general facts of the situation we find ourselves in today – we hope you'll agree. We will discuss the specifics in more detail later on, but because we want to keep the book relatively short and succinct (and therefore inexpensive to print)—and, other than the catalogue of references at the rear that cover Covid-19 related topics in more depth—we will try here to provide a solid *layman's* understanding of the essential basics about the Covid-19 phenomenon as we attempt to get at the real truth.

Because, notwithstanding the various conflicts of interest of the narrative-pushers such as government-hired PR specialists; the compromised national news outlets; and all of those well-paid, Big Pharma-sponsored fact-checkers and theory debunkers on social media; there are just too many conflicting sources and contradictory claims about specific issues out there, all competing for the public's attention. Getting into those arguments will only distract us from the goal – in fact – the distraction itself could very well be part of the undeclared, hidden agenda. So, we are just going to stick with the provable facts as we forensically explore questions such as the precise origins of the virus, its genetic makeup and the efficacy and reliability of the various vaccines so as to arrive at *the* definitive truth. But we must do so on the basis of a solid understanding of the nature of human society and the structures that hold it together including the characteristics of those who are running the proverbial 'show' because in addition to our own research – and complimenting the same – unsettling new data that surfaced in the past few weeks provides a number of intriguing clues and compelling scenarios to be considered, and some rational, logical conclusions to come to.

*   *   *

Now, any objective exploration of any subject requires a clear grasp of the facts. To be clear; we have only produced this book as a personal, moral response to a collective moral dilemma. As a teacher by vocation and a forensic truth-seeker by inclination I feel an obligation to share what I know (vs what I merely 'believe') in service to my fellow man – (and to my fellow women of course, and to any species in-between). To a certain extent I am asking the reader to trust what I am saying on the combined basis of logic and intuition and upon the extensive research that was done before presuming to produce a book that aims to provide clarity on the subject, rather than cause even more contention or confusion. It is assumed of course (and indeed strongly encouraged) that the reader will do a simple online search if any piece of information does not seem quite right to them.

Having said all this however, and before we move to Chapter Two, we need to line up a couple of our ducks so-to-speak, beginning with the observation that we are already, collectively, waist-deep into this Covid-19 crisis and no matter what the future holds, that things will never be quite the same again. Just as during the outbreak of World War One for example, that culminated with the eerily-similar and equally-suspicious so-called 'Spanish Flu' pandemic that reportedly killed up to 100 million people, and even as modern historians continue to struggle with some of the alleged wartime 'facts'; the

17

events and personalities, and the social movements and national perspectives that triggered that great global tragedy; the reality is that all of those reported 'facts' – even if they were fully known and understood by the people of the day – would have been of little consolation to millions of grieving wives and mothers as they spent the rest of their lives in mourning and in abject dismay at man's insane capacity for inhumanity and bloodshed.[i]

And so it is with this Covid-19 phenomenon. Untold damage is being done – and will continue to be done – as a result of its very existence. Whether by cause, by effect or by consequence – or indeed even by necessity – the fact of the matter is that our world has already radically changed and it matters less at this point as to *how* it came about. The more important point right now is in accepting that we are already in a different reality and there's simply no going back. At least, not to any 'normal' as it was before. The Covid dead can NOT be resurrected by any new change in the policies or decisions that sent them to their lonely graves, and the fundamental rights and authorities that we have surrendered to the State – or more accurately perhaps – that have since been denied to us, and/or have been stolen from us, will be very hard to win back.

These are simple but tragic realities that remain unaffected by each of our personal beliefs or by our individual positions on 'the Covid-19 phenomenon'. Whether sceptical or credulous, whether trusting of the authorities or not, these are the sad and disturbing facts and there's no escaping them. People are dying. Life as we know it is changing and, judging by the evidence thus far, it is NOT changing for the better. We need to acknowledge therefore that a great and terrible thing is happening to us right now, and, if we are to play any active parts in our own immediate destinies, then our first and most pressing priority is to understand who exactly is running the show and where are we getting our information from?

Because if we are to truly understand what is happening – and what we are going to do about it – then all of the important questions on the table need to be asked in context of the state of human society today, right here and right now.

If we think of the Covid-19 phenomenon as a prospective crime scene (because that may yet prove to be the case) then that is a good place to start, especially as we now have additional evidence after-the-fact to help us revisit the original events and subsequent developments to see what logical conclusion we may arrive at and where any proverbial 'smoking guns' may be hidden. Clear, forensic thinking in other words, based upon ALL of the facts and evidence now in play. To this end, and for the singular purpose of establishing a common template through which we can ascertain 'the truth' (or at least as close as we can get to it) we will use the tried-and-tested "5W+1H" formula to uncover the facts and the truths behind the Covid-19 phenomenon; i.e. "who, what, why, when, where and how?" Arguably, this is where any rational examination of the crisis should have begun in the very first place; another incongruous anomaly in an ever-lengthening catalogue of inconsistencies, yet to be explored.

To be clear; this is NOT the time for ignorance or apathy. We cannot just sit on the fence and wait to see what transpires, because inaction in circumstances like this equals complicity with whatever evils may be coming down the tracks – as history has so painfully taught us. The default mode of power (as we shall see) is to seek ever more power, and we may be living through the biggest power-grab in history. The time of coasting along in blissful ignorance while 'they' (the authorities) do all of the important thinking and decision-making for us, is over. It is over because no-one is coasting along any more, even if they believe themselves to be. In fact, there is no proverbial 'fence' to sit on any more. You either get with the program or, you start questioning it, because the longer this goes on, the smaller the window of opportunity for us to change course; to make different choices; or to protect ourselves and our communities from the tyranny that is emerging.

And whether that tyranny is a response to a genuine health crisis –

or whether it is born of the excesses of socio-economic globalism and political corruption – the result is the same; a disempowered and disenfranchised population reduced to being mere objects of obedience, with no real say in their own futures.

Being truly human means more than this. It has to. Otherwise we are no better than the beasts of the field, soul-less and submissive. And before our animal-loving friends take any umbrage at this apparent slight to some of man's best friends let me just remind everyone of the notion that 'something special' sets us humans apart, and that special 'something' – our unique ability to self-reflect and consider matters of truth or morality for example by engaging our minds and consciences – has a very big part to play in how we deal with this Covid-19 phenomenon.

In fact, a clear, practical and realistic understanding of the psycho-spiritual notions of soul, and of conscience and of human morality in context of the Covid-19 phenomenon are key to seeing both the great and enduring threat that we are facing, as well as the emerging opportunity for humanity as a unified collective to finally realise what it means to be *truly* and *fully* human, and to then act upon that awareness while there is still time to do so.

This awareness includes our responsibilities to ourselves and to others – including future generations – to seek out what is genuinely best for us all, and perhaps more importantly, NOT to participate in the Great Lie – either through ignorance, complicity, conditioning or plain old mule-headedness – either unwittingly or otherwise:

*   *   *

"For what does it profit a man, if he gains the whole world, but loses his soul?"

Matthew 16:26.

CHAPTER TWO

'THE ESTABLISHMENT'

SO – WHO'S RUNNING THE SHOW?

Firstly, and to put this whole discussion into context, there is the overarching fact that practically all of the 'public data' about SARS-CoV-2 (the virus) and Covid-19 (the consequent disease) is coming from 'official sources' channelled via a largely-supportive and compliant mainstream media, and that the public's dependence on that information and each person's own level of trust in the same, are at the very core of this contentious debate. Because somewhat obviously, if those official sources have any alternative agenda other than delivering the unvarnished truth to us, then absolutely, we need to be asking whatever questions we need to, to establish the truth and ascertain the realities behind the whole Covid-19 phenomenon as it relates to – and is being experienced by – us ordinary folk in society.

The Establishment: For the sake of clarity, when we reference 'the establishment' in this book we are referring to that group of people in positions of power, influence and authority worldwide who are united in their common goals to procure ever more status, wealth and power for themselves, and who do so largely through diverse deceptions and coercive dishonesties at the collective expense of the rest of us. Variously known as 'the Cabal' or 'the elites' or by similar other terms that suggest the existence of powerful, interconnected groups of single-minded individuals whose aim is the exploitation of the masses for the objective of profit. Some are independently wealthy; oligarchs, billionaires, and owners of multinational corporations. Others have inherited political power or have been appointed or elected to high office, and numerous others have forged careers in so-called 'public service' or are otherwise sufficiently well-placed in the public eye (such as media chiefs, community leaders or celebrities) to have been embraced into the clique. In very simple terms, these 'highly-placed' ruthlessly ambitious, but often-amoral people depend; (a) upon the unquestioning compliance of their employees (in corporate settings); (b) upon the ignorance, obedience or submission of the general population (in State-funded agencies); and/or (c) upon the cultural conditioning of the masses to achieve their goals. With the rise of super-rich corporations and high-tech operations embedded on different continents in often-covert partnerships with regional governments, and with all of this being facilitated by mass media and modern digital communications, it is easy to understand how a global network of such like-minded individuals is now very well established.[1]

The Government: Most of us live under the misapprehension that our governments actually run our countries, and that they work for us. This is NOT the case – at least not any more. There was a time of course when kings and emperors for example held central control over particular territories or peoples, and any laws they issued directly affected those populations. But autocratic, repressive systems like those were so inclined to tyranny that the people eventually got fed up with them, and so, after a few heads had rolled and necks had been stretched, modern democracy was born. The original idea was that people would actually 'govern themselves' by electing public representatives who would speak on the peoples' behalf and bring in laws and policies to that specific end. Presumably, before politics became a money game, these early representatives were chosen by virtue of their personal qualities; their integrity and wisdom, and 'the will of the people' was thus organised and condensed into a practical form of administration that we now call 'modern democracy'.

But a word of caution needs to be inserted here, because even if the *ideal* of democracy is indeed a theoretical improvement upon overt tyranny, we must also acknowledge the historical fact that no sooner

---

[1] See, *"The Establishment – and how they get away with it"* by Owen Jones.

are these new-and-improved political models of supposedly 'participatory governance' set up, that moves are immediately afoot amongst the ruthless and the ambitious to re-establish their own particular tyrannies within that democracy – either by stealth or deception – because no self-respecting tyrant is ever going to allow a largely unaware peasantry to become an impediment to their own grandiose plans, nor to those of the current establishment. By sheer force of numbers the peasantry may be able to force the Cabal – for a time – to deploy democratic terminologies and engage in what *appears* on the surface to be democratic debates and processes. But ultimately, as long as we (unaware peasants) continue to believe that we can replace our own personal moral authority with some sort of collective political institution – and as long as we allow ruthless, ambitious and deceitful profiteers to ascend to positions of power in those institutions – then the way is open for tyranny to re-emerge and flourish in one form or another, just as we are witnessing today as we watch our purported 'modern liberal democracies' collapse chaotically into heavy-handed tyrannies during this Covid-19 crisis.

Dramatic examples of democracy-replacing-autocratic-tyranny in relatively recent history would include those that arose out of the French and American Revolutions, with the U.S's *"Declaration of Independence"* in 1776 being the inspiration for scores of similar populist upheavals around the globe, including France's 1789 *"Declaration of the Rights of Man and of the Citizen"* which arguably, is the precursor of all subsequent human rights movements. Thus 'sovereignty' in the sense of dominion, or control or authority for one's own governance was transferred from one autocratic sovereign (the King or Queen) to the individual acting in a democratic collective. Thus the ideology behind those two history-changing declarations – as well as the general acceptance amongst the world's population of the primacy of the democratic model of government (vs autocratic rule) – marked a radical changing point in human social history, and it is fair to say that most people today accept and believe that democracy is probably the best system of government that humans have come up with so far. Whether we are right though, remains to be seen.

You see, bad old habits are persistent, and it didn't take too long for emerging democracies to produce their own internal autocracies complete with their respective 'kings-and-princes' and entitled ruling classes (in the form of burgeoning political parties for example) who then competed with each other for control of the masses, and of course, for access to of all of the riches, and resources and powers that come with that control. The only difference between the Royal Courts and the dictators of old – and our modern democracies, is that 'they' (the establishment) now provide us with a selection of insider candidates from whom to choose our next rulers. And who is to say that even *that* most precious bastion of democracy – the individual vote – is NOT being directly interfered with, in addition to all of the political propaganda and half-truths that the public are bombarded with at election time? Shameless propaganda that is ultimately, funded by the taxpayer. Thus, we effectively pay the political establishment to lie to us, and then, we reward the best liars with power. Yet still we wonder at why all of this deception and dishonesty in politics? So although we still call our supposed modern liberal democracies "democratic" surely, isn't the term 'modern democracy' a mere euphemism for, *"whatever process or pretence we (the establishment) have to suffer to ensure the propagation of the status quo"?* Because ever since the first banking crisis it is evident that the interests of the Banks and of 'big money' trump those of the ordinary people. They gambled – but we lost! And I don't recall us having any vote on that bailout decision. So, what happened to our democratic process when we needed it most? Because it does rather look now as if our governments have been working for someone else - no?

Theoretically of course, the three branches of government are *supposed* to be playing interdependent roles each with separate, distinct powers in variously making, applying and interpreting the laws. This 'separation of powers' doctrine was designed to eliminate insider corruption (in theory at least) and thereby guarantee our democracies and the Rule of Law. These three branches are; (i) the Legislature (Parliament); (ii) the Executive (the Head of Government and the Cabinet); and (iii) the Judiciary. Apart from all of the protections written into the Constitution to protect the peoples' interests, there are also flocks of civil servants, statutory oversight bodies, police and public office holders to keep a strict eye on things. But as if all of this purported 'independence and oversight' wasn't enough, we also have the mainstream media acting as an unofficial 'Fourth Estate' – a supposed 4th and *truly* independent, autonomous and self-funded 'pillar of democracy' that reports uncompromisingly on the issues of the day, including exposing errant authority figures and holding them to account (at least in the public eye) thereby ensuring openness, transparency and accountability amongst those who would call the shots; ethical standards that are fundamental to any healthy society that values truth, justice and transparency.

The Media: Unfortunately, the mainstream media in general ('MSM') can no longer be described as truly 'independent' simply because the major media corporations have multiple interconnected outlets – all individually branded to give the *appearance* of independence – but which are owned and controlled by a select few, highly-placed members of the global-corporate establishment. These individuals who control large swathes of newspapers, magazines, radio and TV channels and other news and entertainment outlets we usually refer to as 'media barons' or news tycoons, magnates or moguls etc., and most of them are highly-driven entrepreneurs with a singular eye on profit. In other words, given a choice between two stories for the front page – and regardless of the social, moral or philosophical importance of either, they will opt for that story which will sell more newspapers – or attract more advertising revenue. It's really as simple as that. News stories are selected not on the basis of their social or moral importance to you or me – the end-product consumers – but upon what brings in the moolah either directly or indirectly to the franchise. *Directly* (as we've seen) through newspaper sales or advertising, and *indirectly* through the protection

25

of affiliated interests or associated businesses who contribute in some way to the proverbial 'bottom line'. Thus, for example, a story that may be embarrassing to a highly-placed politician who also has a role in sending revenue-generating official stories or notices to that news outlet will not get aired, whilst the very same story about a mere commoner being prosecuted in the Courts for the very same offence, will. Because after all, in the end it's all about the bottom line – right?

Likewise with the influence of 'Big Pharma' for example. If someone were to inform you that the pharmaceutical industry is *the* highest funder of advertising revenue in the Unites States for example, and that all but one of the major news corporations share a senior Board Member with one of Big Pharma's giants? Would that fact raise any natural concerns about the integrity or quality of the news when such massive profits could, possibly, be earned by deceiving the public? In particular context of the Covid-19 phenomenon, it should really trouble us that those same news outlets received over five billion dollars from the likes of Pfizer, Moderna, Merck, Johnson & Johnson *et al*, who, quite coincidentally of course, are profiting massively on the back of this mainstream-media declared 'global health crisis'. And before all of the innocents amongst us rebut any such scenario with a robust declaration of faith in 'the statutory authorities' who will (they righteously insist) most assuredly address any such compromised relationships between big business and the media by investigating 'in the public interest' what is in effect, massive-and-insidious white-collar crime; we should be aware of the officially-undeclared policy of the Irish State Prosecutor's Office for example, NOT to prosecute 'white-collar' or regulatory crime for two stated reasons: (i) because big businesses have deep pockets and can litigate *ad infinitum*, and; (ii) because the government doesn't want the public losing confidence in publicly-funded institutions.[ii]

The appalling irony in their 'official position' seems to have escaped them – or has it? Because they are in fact assisting, facilitating and covering-up corporate crime "in the name of the people"! In other words, even if it is obvious that crimes are being committed on an industrial scale by corporate dirt-bags sitting in plush offices, and even if those crimes are costing the taxpayer many multiples of whatever 'ordinary' crime is going on in Ireland (i.e. "crime in the streets vs crime in the suites") then the DPP's Office is unlikely to prosecute, so as not to unsettle the perceptions of a naïve and trusting Irish public. Inasmuch as those criminally-compromised institutions or businesses are integrated within the establishment there must also be some suspicion at the very least that the third unspoken reason for such an obviously-unjust stance by the nation's prosecuting authorities is plainly-and-simply to protect the interests of the Cabal. A voracious cabal (more on this later) who fully knows, accepts and understands that in order to foist their various schemes on an unwitting public under the guise of 'lawful propriety' that they need the compliance and protection of the national authorities to secure government contracts at 'highly favourable terms' for example; or, to continue the petty frauds and larcenies (largely via the banks and vulture funds) that destroy ordinary peoples' lives. Remember when we were told that the (criminal) banks were, "too big to fail" and that we taxpayers would now have to bear the cost of *their* reckless speculations and greed-mongering? Add to this the mass plundering of the population through politically-sanctioned appropriations of national resources and the deployment of clever tax avoidance schemes, plus the all-too-cosy relationships between big business, the media and the political world... and the true picture starts coming into focus.

When we consider all of this in light of the fact that Ireland (for example) is by far the world leader in the dark art of 'shadow banking'; that we carry over €200 billion in debt and yet still 'own' over $300 billion of USA debt, making tiny little Ireland the third largest holder of US debt after China and Japan; and if we then add in the fact that Ireland hosts the European headquarters of Google, Facebook, Apple and Microsoft – all very big players in the Covid-19 phenomenon; not to mention the ease with which the European Union coerced a supposedly 'sovereign' Ireland into re-voting for the

Nice and Lisbon Treaties a second time so as to keep the EU project on track during a political campaign that made a total nonsense of the notions of sovereignty, of democracy, or even of basic honesty. It was little old Ireland too who was saddled with 42% of the EU's foreign banking crisis debt despite comprising only 0.9% of the EU population and making up only 1.2% of the European Union economy – a fact that should leave no-one in any doubt of the truly 'special' place that Ireland holds in the global Cabal, especially in terms of its abject willingness and ability to comply unquestioningly with unjust and punitive diktats at the great expense of the Irish people. This is just one apt example of how-and-why 'big business' in its various forms, and their often-secretive partnerships with national authorities and the mainstream media needs to be very carefully watched indeed.

Amongst some of the larger media giants that have multiple supposedly 'independent' outlets in the USA and in Europe for example, we have:

- **AT & T** incorporating: Time Warner; CNN; Warner Brothers; CW; TBS; TNT; DC; HBO; Cartoon Network.

- **Walt Disney** incorporating: ABC; Fox; 21st Century Fox; Pixar; marvel Studios; National Geographic; ESPN.

- **Comcast** incorporating: NBC; Sky; Telemundo.

- **Viacom** incorporating: Paramount Pictures; MTV; Nickelodeon; Black Entertainment.

- **Mediahuis** which owns over 100 major newspapers, as well as TV and Radio stations in Holland and Belgium including the Independent News & Media Group in Ireland.

And, just to make the point without listing every single media tycoon out there, the list of media outlets and new publications owned by **Rupert Murdoch** for example included: News Corporation Holdings with three national newspapers in the UK; almost 150 publications in Australia; the New York Post and Community Newspaper Group in the United States; The Wall Street Journal and related publications in the U.S., Europe and Asia; as well as the Dow Jones information services. Whilst Murdoch's broadcasting businesses included the FOX Broadcasting Company; the 27 stations in the Fox Television Stations group and various television operations throughout the world. Other outlets produced and licensed programming for cable and satellite platforms in the U.S and Asia, including the FOX News Channel and FOX Business Network, FX and STAR. News Corporation also owned Italy's most popular pay-TV company, SKY Italia. The company also had significant holdings in British Sky Broadcasting, Germany's Sky Deutschland; Asia's TATA SKY and FOXTEL in Australia and New Zealand. Rupert Murdoch also controlled movie production and distribution through Fox Filmed Entertainment and Twentieth Century Fox Film, as well as television productions through 20th Century Fox Television and other TV studios. Other assets included "next generation" media properties including an online video joint venture with NBC Universal and Disney; and News Outdoor, an outdoor advertising company.[iii] And that was ten years ago. Lord knows what the list looks like now. Anyway, the point being made is that certain people with massive wealth and influence are, if they so wish, absolutely in a position to dictate what we see, read or listen to as 'news'. Furthermore, if they are wedded to the political establishment – or to the Pharmaceutical industry for instance – either for ideological or profit-related reasons, then only a fool would not be suspicious about mainstream media being used by these hidden actors, as propaganda.

In the event that the reader feels we are being a bit harsh on the 'independent mainstream media' or are unfairly tarring all of its owners and senior editors with the same corporate brush, we must of course acknowledge that there simply *must* be some decent, courageous journalists of integrity out there. Unfortunately, we just can't find them. At least, not any that still work for the major news networks, because those few journalists of integrity who have tried to raise public interest stories that might embarrass highly-placed figures or are otherwise 'inconvenient' to the Cabal, soon get their wings clipped. When an award-winning reporter stumbled upon high-level corruption at Garda Headquarters in 2014 for example she soon lost her job. The fact that her new boss was previously the Editor of the Garda Magazine hardly raised an eyebrow in establishment circles, and the damning facts that; (a) the newspaper had to pay out, "an undisclosed (but substantial) sum in damages", and (b) that the reporter's previous and subsequent work had indeed uncovered some serious establishment scandals, are now merely footnotes in a Wikipedia article that is almost entirely devoted to the derogatory 'rebranding' of someone who continues her work as an influential, outspoken anti-corruption activist and as a genuinely independent journalist.

Other prominent reporters who have shown interest in controversial State-linked stories soon get put in their place too. One senior journalist (who we won't name for his own protection) told us that his interest in the Ian Bailey case (which is one of Ireland's longest-running crime scandals) nearly cost him his career. His investigation into how and why the Gardaí were paying off a local informer with illicit drugs from the evidence room – or more alarmingly – into how a five-bar farmer's gate complete with bloody handprints from the murder scene had simply disappeared from evidence, was clearly a valid story to pursue. "I could have torn up my contact book after that" he told us. Controversy still simmers over a case which, to those of us who know the inner workings of the State apparatus here in Ireland, is such a typical example of how 'the system' will twist, bend and break any rule it wishes so as to achieve the required result. Crucial evidence going missing? Ah well, these things happen.. People being set-up and framed by the police? Witnesses being variously threatened, coerced or bribed with drugs? It's all just business-as-usual I'm afraid in diddle-de-dee Ireland where the truth means nothing. The law means nothing, and ultimately people's lives mean nothing – not when the alternative might cause a bit of embarrassment to the Cabal.

In our recent publication, *"Criminality in the Irish Courts and the*

Absence of the Rule of Law" there is a story beginning on page 79 that makes reference to a serious criminal assault and the subsequent murder of one of the perpetrators which occurred in 2010. The whole extended story, including layer-upon-layer of State-orchestrated obfuscation by gardaí, by the Irish media and by all of the so-called 'statutory oversight bodies' including the Courts was purposefully suppressed and ignored by the Irish establishment in favour of their own contrived 'official narrative'. The story (of that story) was covered by Berlin's Cicero Magazine in 2017. Three years later and only a few short months ago, evidence came to light that one of the individuals connected to that assault and murder (who was never even interviewed by the police) had hired a hit-man in the UK to kill a family member in 2020. We kept all of the evidence; the emails, the recorded phone calls, texts and witness statements and sent the contact details of the lead witness to the UK authorities, advising that once they started their investigation, we would forward the details of the hit-man to them. We also told the UK Police that it now appeared that the Irish Gardaí were heavily implicated in covering-up (at the very least) the 2010 crimes. This last little titbit seems to have been the reason why the UK establishment – all the way to Priti Patel at the Home Office, including two regional Police Authorities and five MP's spent the next six months scrambling to avoid doing any proper investigation, and then desperately covering up the fact. The witness too fell mysteriously silent after being assaulted by 'three unknown assailants' just a week after we first alerted the UK Police, and even when we repeatedly asked the pointed question over a full 6-month period, *"Why doesn't ANYONE want to know the name of the hit-man?"* ..we still got no takers. So finally, we went to the top UK newspapers. To 150 of them. The only response we received was from Channel 4 News, who advised us that we had used the wrong email address. We have heard nothing back from any of them since.

When you can literally report a murder and a conspiracy to murder – and then inform 150 supposedly-autonomous news outlets that you have a rock-solid story of police cover-ups and criminal corruption and you get no takers at all? Well, if that doesn't point to the role the modern mainstream media plays in protecting the status quo, then maybe nothing will. The story would undoubtedly have made a small fortune for whichever news agency covered it. But apparently someone somewhere with sufficient clout over so many supposedly 'independent' news outlets decided that it was more beneficial in the long run NOT to publish a story that implicated so many highly-placed members of the establishment in outright criminal activity.

So, one might ask, how do we account for the occasional big scandal that threatens the establishment that *does* get exposed in the mainstream news – such as the Edward Snowdon / Julian Assange / Bradley Manning and Wikileaks stories for example? Well, all I can say is watch the movie, and note how terrified the reporters were at the prospect of being caught. Then look what happened to Snowdon, Manning and Assange since – simply because they told the unvarnished truth. Once stories like that get 'out' it is often then about damage control. Better of course (for the establishment) if the story doesn't get out at all, but if it somehow does, then the usual crafted denials, misdirections and cover-ups come into play. This applies to several aspects of the Covid-19 story as well – as we shall soon see. The story is suppressed through systemic lies and denials or, by targeting the source directly via intimidation and threats; by public vilifications; through gaslighting and character assassination; and/or by questionable or trumped-up criminal charges. If all else fails, then that's when the stonewalling comes into play; the last, cowardly retreat of career liars. When 'they' hold all the power and pull all the strings, this really isn't so difficult to achieve. And, once the initial storm has passed, they then rely on 'the fade factor' – the anticipated waning of interest by the public, until it's "back to business as usual".

The fact of the matter is that for years now, we (the *Integrity Ireland Association*) have been copying-in the mainstream media to correspondence and court cases, to criminal complaints and petitions, and to proofs of all manner of outrageous abuses of our fundamental rights by government agencies in particular. Even the so-called 'statutory oversight bodies' and purportedly 'independent' human rights agencies have been exposed in their various often-undeclared ties to government. Some of them are even funded by government bodies raising the all-too-obvious question of how then can they be truly 'independent' or claim there is no conflict of interest when they invariably reject our evidence or refuse to act upon it? It is hard to get one's head around it sometimes, but we now have mountains of evidence that proves (in Ireland and the UK at least) that the default position of practically all of the institutions of State (including the MSM Fourth Estate) who are *supposed* to be working in the public interest, is to protect the establishment and defend the great lie.

Some years ago – before we really understood how the system works – we alerted Ireland's most prominent crime correspondent that we were having difficulty getting the Gardaí, the Garda Ombudsman and the DPP's Office to respond to allegations and proofs of serious political corruption. Incredibly, the reporter responded by threatening us with defamation proceedings. We countered, asking him why RTE's chief crime correspondent (himself) wasn't apparently interested in even seeing the evidence? We never heard from him again. He is still RTE's main crime correspondent, broadcasting nightly – with an appropriately-serious look on his face – as he reports on all of the crimes being committed by the peasantry. He also remains one of only two reporters who have direct access to the Garda Press Office. Not that we're suggesting of course that it might be more transparent and less suspicious (of any conflict of interest) if we didn't have reporters in those roles whose daddies were Garda Superintendents.

Thus, the 'news' we get is not so much informative of the *real* state of affairs, but is instead a highly-sanitised and propagandised version of 'reality' that only serves 'their' interests. Like so many corporate entities active today, the operational focus in the mainstream media has shifted from delivering a genuine service or news product to the people – to simply maintaining the external *pretence* of truthful journalism whilst generating more profits and/or protecting the often-insidious interests of the status quo. Sometimes, as was recently decided in a US Courtroom, the newsreaders fully *know* that they are misrepresenting the facts, and are deliberately misleading the public. Their defence to a claim of deception by intent and even personal defamation was simply that they weren't actually delivering 'news' *per se*, but that their news programme was just another form of common entertainment – one where the newsreader, as an actor, had a certain amount of 'creative licence' no less. A licence to lie in other words. And believe it or not, the ruling of the Court was that "Yes! ...viewers *should* have known that 'the news' in this case was merely inappropriately-branded 'entertainment'." So much for journalistic integrity and for the wisdom of the Courts eh?

Teaching a University course in communications some years back, I had occasion to explore the various methods of communication and the skills and methods required to achieve the stated objective. Was the objective to inform, to teach, to direct, to conform, to enquire or to entertain for example? From casual conversation to issuing military orders; from telling a children's story to reading a scientific report; or from delivering stand-up comedy to reading the news etc., etc. Each discipline requires a very different approach because each has a very different outcome – right? News becoming mere entertainment that contains lies and defamations is clearly not real news any more. So why do they persist in calling it that – and where on earth did they find that stupid judge? (And no Judge, it's not defamatory if it's true).

So there we have it. Should we still have confidence in news providers in the knowledge that newsreaders are just actors and entertainers? Script-readers and propaganda deliverers who are just 'doing what they're told' by the corporation or the government? And it is of course only natural, and is most certainly a prominent feature of corporate structure today (if you want to get promoted that is) that you do what the boss wants, and if the boss wants you to cover this, that-or-the-other story from this, that-or-the-other perspective, or, if he wants you to drop the issue altogether or just make up a few entertaining lies, or some false facts or defamatory smears, then that's exactly what the obedient employee-reporter-actor-journalist does. Because again, this is 'a business' after all, and profit is king.

'Deep State' media control. Worth a brief mention here is the CIA's Operation Mockingbird which was a project that began in the 1950's whereby scores of CIA actors were positioned as news anchors, journalists and even as clergy in local churches so as to influence and yes, control, public perceptions. Arguably, these were the first orchestrated efforts at producing 'fake news' on a scale that would effectively shape American public opinion and from there go on to influence world opinion with all of the political and commercial decisions that would flow from that; based on carefully-crafted lies.

> *"Since the 1950s, the CIA started recruiting journalists, editors, and students in order to write and promulgate false stories. The CIA's stories were entirely propaganda and their employees were paid huge salaries in order to promote such fake news. Essentially, the CIA managed to control both national and international newspapers through a bribe."* [iv]

Linked to other initiatives such as the CIA's undeclared ownership of Crypto AG, the Swiss company that has provided at least 120 countries with specially-encrypted technology since the 1970's for sensitive or secret internal communications, it is not such a leap of cognition to picture a scenario where the same State-controlled (or Deep State) mechanisms remain in place today, only in far more sophisticated and comprehensive formats. [v]

State-sponsored news and entertainment outlets such as the UK's BBC (or RTE in Ireland) are *supposed* to be different to commercial media platforms inasmuch as they are funded by the taxpayer with a mandate to provide truly independent 'public interest' programming. When the BBC started out for example that's largely what they did and as a result, they commanded an unrivalled worldwide respect. But any media outlet that is directly married to the government is always going to come under political pressure to deliver what the government wants as opposed to what is genuinely in the public interest. It may take some time and organisation and lots of lobbying and horse-dealing before they eventually get their jockeys in place, but where there's a will there's a way, and the longer that marriage – especially if the same political party is returned to power – the surer   the political takeover. Ireland's RTE exemplifies the almost total corruption of its original mandate, with insider political appointments to key senior positions and with the government using national radio and TV – quite unashamedly – as their propaganda arm. It's not even a secret anymore. RTE no longer covers independent political candidates for example as they used to do (unless you are hitched to the big boys) and we were personally informed by a reporter working for a prominent radio station that he had received direct instructions from the government of the day, that he was NOT to cover *Integrity Ireland* or Stephen Manning during the 2016 elections. (Ironically, we were using the occasion to put the spotlight on political corruption).

Even more ironic perhaps is the fact that whilst running in 2016 and 2020 (for the same pro-justice reasons) under the combined slogans of; *"Lets Shake Up The Establishment"*, and *"Vote Manning, NOT a Politician"*, and even under the title of *"Honest Outlaw"* due to the fact that we have been denied access to justice across the board; that at the first election some 25% of our votes went mysteriously missing between the tally and the 1st official count, and then in 2020, after formally warning the Returning Officer (whom we had previously raised criminal prosecutions against in his role as County Registrar) that "there would be serious consequences" if the same thing happened again.. we discovered that this time, over 56% of the vote disappeared without trace or explanation under his watch. Naturally, we immediately raised a complaint and demanded a recount, and, when we were unlawfully ignored, we lodged a criminal complaint with the Gardaí. When that went inexplicably dead in the water, we followed up by alerting the Houses of the

Oireachtas (the Irish Parliament) and made an application to the Courts. All have been completely, utterly and unlawfully ignored! One would have thought that in a modern democratic republic like Ireland where the Rule of Law and the integrity of the vote reign paramount in our continued membership of the EU and the UN, that someone involved with law enforcement or indeed someone involved in the political process would be sufficiently alarmed at these unlawful and undemocratic events to raise even one lazy eyebrow in enquiry? But no? Not the police, not the politicians, and not even the local media. The reason? Well, quite simply any proper response would be too embarrassing to the establishment – and worse still, it just might expose the great lie.

Another recent development is the hiring of successful MSM reporters as political spin doctors and speech writers for sitting politicians. With exorbitant salaries being paid out of the public purse to these 'special personal assistants' or 'media consultants' these job offers are seen as a promotion from the corporate MSM; a reward if you like for not rocking the boat too much in their former careers. Thus the journalist's creed of, "public trust, truthfulness, fairness, integrity, independence, and accountability" is exchanged for the great lie. Even the most undiscriminating prostitute would blush at the brazen duplicity and hypocrisy of it all. Although those noble principles of 'truth, justice and transparency' continue to be widely broadcast as the foundations of our various democracies and as the purpose for all of our laws, the reality is that these ideals have long since been abandoned by the statutory authorities and by their corporate handlers in favour of an undeclared substitute philosophy that has far more in common with the tyrannies of old than with any genuine notion of people-centred democracy.

Social Media & Big Tech: The cloaked takeover of our political systems by multinational corporations will be covered later on, but the more obvious example of the loss of operational independence remains in this takeover of the mainstream media by super-wealthy individual members of the establishment – or by clandestine dark actors of the Deep State as we have just seen. And this has led, somewhat ironically, to the rise of social media as an alternative news source. But there are two major problems with the explosion of social media. Firstly, and somewhat obviously, because practically anyone can launch themselves as a 'news source' online without any proper training (and therefore without the requisite discipline to research the subject properly) then unless a person or website has built up its own credibility over time there is little we can do – other than another cursory search online – to verify any particular story. But then again, how is this credibility-proofing any different from those compromised MSM news and entertainment sources we have just been discussing?

The second big problem is that most of the major social media giants (Facebook, Twitter, YouTube, Google, Instagram) have now weighed in as self-appointed moderators and fact-checkers and censurers of the materials they are hosting, and they are clearly on the side of the establishment. Or, at the very least, they are 'holding the official line' either willingly, or in ignorance of the various hidden agendas. Now we have an information 'war of attrition' going on where truth, transparency, and freedom of speech have become the first casualties. Indeed, it could arguably be said that the compliance of so-called Big Tech giants who have so much direct control over our communications (Apple, Microsoft, Google, Twitter, Amazon *et al*) and who have collated so much of our personal data between them, are an integral part of the proposed 'New World Order' being so vigorously pushed by the establishment. Thus, like so many of their corporate counterparts, the major social media platforms too have become a not-so-covert extension of the establishment itself whilst maintaining the external *pretence* of independence, and of open-and-transparent service to the public. We will deal with the phenomenon of burgeoning fact-checking agencies in more detail in Part Two, but I suppose the point we are making in context of the Covid-19 official narrative in the MSM, is can we really trust what we are being told?

Likewise with the three branches of government; those so-called 'pillars of democracy' and bastions of purported "independence and public accountability". These too have long since been infiltrated and usurped by agents of the establishment through the slow, inevitable corruptions that accompany the rise to power, especially when the public is not really paying attention, or is distracted elsewhere. There are many immediate rewards for those who are willing to sell their souls. Money as they say, is the root of all evil. Perhaps nowhere is this more evident than in the treachery of duplicitous politicians and the betrayal of our hard-won democracies. Yet still, we continue to re-elect these career rogues. Mesmerized apparently, by the great lie.

And what, one might ask is that great lie then? Well, in a nutshell, the exposure of 'The Great Lie' is a central theme of this book, but there is considerable difficulty summing it up in one short sound-bite because the lie is so great and so perennial – and is so much a part of modern life – that just like the old saying goes; "One can't see the woods for the trees" and so most of us will struggle to grasp its very existence. This is what is meant by the term 'cognitive dissonance'. The inability to perceive or grasp something that is staring us right in the face because it challenges our existing beliefs or conditioning.

Secondly, there's the disturbing fact that we all, in one form or another, are playing our individual parts in that great lie, albeit mostly subconsciously, unwittingly and/or unwillingly – (although some do actively nurture and embrace it) – and disturbing concepts like these do not make for pleasant or comforting reading. But if we are pressed to sum up the Great Lie in succinct but in arguably somewhat obscure terms, we could accurately say that, "The great lie is – that there is no Great Lie!" American scholar Naom Chomsky put the concept neatly when he said: *"The problem is that people are unaware. What's more, they are unaware that they are unaware."*

And what is it then, that we are all unaware of then Mr Chomsky? Well, I would repeat again of course that it is, 'the Great Lie' which in turn raises the all-too obvious question of whether or not we want to *know* what that great lie is so that we can escape its clutches, or, are we happy enough to just meander along playing our unwitting part regardless of the devastating consequences for others – as long as we *personally* are not directly or knowingly affected by it. Because as another old saying goes, "Ignorance is bliss" - right?

In other words, are we ready for some uncomfortable truths – really? Or do we still want to maintain the reassuring lie? Because this question will keep resurfacing in context of the current crisis until each of us is forced to choose. "Do I want to know the truth and will I accept the responsibilities and obligations of that knowledge; or, am I happy living with the lie?" Because this time ignorance does NOT equal bliss. There WILL be consequences for facilitating the lie. Just as in any time of danger or of great social upheaval or of war for example, ignorance or apathy or unawareness – especially when coupled with stubborn closed-mindedness – could very well be a death sentence. And this is not being said in jest – nor for dramatic effect. It is simply the real, unvarnished truth.

Very interestingly, this Covid-19 phenomenon may be presenting us with the single best opportunity since the Great Wars of the 20th Century to observe and deconstruct the Great Lie so as to allow truth to re-emerge, with a mind to returning humanity – as individuals and as a collective – back to that original place of truth and self-determination that has always been our birthright.

CHAPTER THREE

CROOKS, COWARDS & FOOLS

AN INVITATION TO TYRANNY

In his insightful treatise entitled, *"The Discourse of Voluntary Servitude"* written nearly 500 years ago, French scholar Étienne de La Boétie summarises the function of centralised governments in respect of the roles of their respective populations (including monarchies, parliamentary democracies, republics and dictatorships) as providing for three classes of people. The first class he calls 'knaves' which is an old term for scoundrels or villains, *"a numerous and active class, who see in the structures of government an opportunity for their own self-aggrandisement or wealth"*. The second class he terms 'dupes'; the vast, foolish mass of humanity who think they are living in 'free' societies simply because they have one vote in millions over what form of tyranny prevails over themselves, or is visited on their behalf, upon others. And the third group interestingly, he leaves unnamed. But somewhat depressingly (because not much seems to have changed since) he describes as (abridged) *"..that class of people who have some appreciation of the evils of government but either do not see a solution, or, if they do, they are not willing to make the personal sacrifices necessary to bring about that change"*.

The whole depressing concept can be summarised by three insightful observations at the front of that book which declare (abridged):

1. That liberty is the natural condition of (the) people.

2. That we have willingly enslaved ourselves to governments.

3. That if things are to change, that we must first acknowledge and understand that there are vast networks of people with a vested interest in maintaining these tyrannies (by any name).

And that was 500 years ago! Those ancient 'networks' for want of a better word comprise 'the establishment' today and all of the 'knaves and dupes' who support them – whether they are aware of it or not. This is not said to alienate or offend those of us who work for the State or for the establishment in some form or another because undoubtedly, many are doing so with the very best of intentions – I did so myself, unwittingly, when I served in the military. But just as any soldier cannot later undo the crimes of their superiors – nor legitimately claim that he was 'just following orders' at the time; (an excuse since outlawed at the WW II Nuremberg trials) – we need to constantly educate ourselves as to the essential dynamics at play and as to what particular role we may each be playing at any given time or place. Otherwise how can we possibly change anything and why would we even want to, if we're not even aware that something's wrong? It matters little what our personal beliefs, perspectives or sincere intentions are if they do not align with the truth. Because as the old proverb goes: *"The road to hell is paved with good intentions!"*

A simpler understanding of Boétie's observations would be that most human institutions and especially those of central governments, exist to serve the interests of the institution and of the establishment, and therefore NOT those of 'The People'. Thus all claims to the contrary are lies. An abundance of knaves are needed however, to mask and disguise those lies, and they too must be compensated. It then falls to the dupes to swallow those lies whilst simultaneously paying for them through the nose in the form of taxes and salaries for all of the knaves, and for the agents of the establishment who run the show.

In this manner, an unhealthy dynamic between those who would abuse our rights (the corporate-political establishment) and between those who would readily give them away (most of the public) has been well established, and almost all hierarchical institutional models in operation today—and especially corporate-capitalist models—have adopted this operational position, mainly because "it works". It only works however, if-and-when sufficient individuals surrender their moral authority (to reason for themselves), in return for personal rewards (such as conveniences, cash, services or career promotions) and/or, in fear of the repercussions or consequences if they do not. And make no mistake, but fear, and the dread of consequences of disobedience or non-compliance are as much a part of the fabric of modern democratic institutions as they ever were in the tyrannies of old. The only real difference is that the ancient tyrants were, by comparison, relatively honest in their oppressions and in their persecutions. Vlad the Impaler for example, made no secret of the fact that he expected to be obeyed without question.

Today we are told that we ARE allowed to question our democratic rulers; that they ARE subject to the law, and that they ARE ultimately accountable to us, The People. Really? Perhaps those of you who still believe this should try raising an issue 'through official channels' sometime. Because, depending on the importance or the pertinence of the question (in relation to the protection of the establishment) you will, at best, find yourself being pin-balled from government-department-to-government-department; from one State agency to another; and up-and-down the never-ending hierarchies of civil and public servants until you lose the will to live.

It is quite remarkable in fact how well the establishment has insulated itself from genuine, democratic scrutiny chiefly through tiers of unquestioning, compliant civil servants whose only concept of 'service' is to themselves, their own venal ambitions, and to the despots who have the power to reward them for their moral abeyance. Thus if you persist in your questioning you will enjoy that unsavoury experience that is usually reserved for whistleblowers and other such annoying 'persons of conscience' and will get firsthand experience of how the State deals with moral irritants. The suite of tried-and-tested 'punitive repercussions' then gets rolled out, leaving the bewildered target asking that essential question of themselves, *"Oh, what's the point?"* and/or *"Is it really worth all this hassle?"* And then, quite understandably, they give up the struggle for the rights, the responses and the respect that we are all entitled to! And this, my friends, is how 'they' do it – and how-and-why they manage to keep on doing it – with arrogance and impunity!

When any such 'punitive repercussions' are dealt out by authorities in response to awkward questions, to political resistance or to the expression of ideological alternatives for example – and especially when this is done overtly, aggressively or with menaces – we quite rightly call it 'political oppression' and we point vigorously at tyrannies like North Korea as examples. But what about the tyranny of deception, of greed and exploitation (of other nations in particular) that has marked the 'progress' of the western world? How can we rest easy knowing that our modern democracies were built on the backs of slaves and at the great and continued expense of Third World countries, something that continues today; a crime against humanity that we are all – whether we are aware of it or not – at least somewhat complicit in?

What if we were to discover that despite our apparent evolution through history from slavery-based societies, through feudalism (in Europe) to the rise of so-called 'modern democracies'; that each such historical development has been quickly adapted to serve the interests of the establishment? Indeed, it could reasonably be argued that none of these socio-political developments could, or would have occurred, if they were not supported by the bulk of the establishment in whatever particular form that establishment took at the time. Even in the case of successive revolutions and wars – and even in the case where one type of establishment such as the French Monarchy for example, got replaced by a Revolutionary Government, which in turn was ousted by Napoleon's French Consulate; we can be sure that 'hidden hands' were at play behind all of these power moves. 'Hidden' being the key operative word here, because it usually doesn't suit the interests of the truly powerful to expose themselves to scrutiny or to accountability. That privilege is usually left to their sock-puppets – to the well-known, public-political faces of change.

Probably the first, best example of 'the establishment' in action in similar form to that of modern multinationals today was the East India Company – Britain's commercial exploitation of India, backed up by political and military institutions, which, by the mid 1700's accounted for half of the world's trade. Colonialism at its very best in other words – only in a 'strictly business' format. Coinciding with the rise of Freemasonry as another pillar of the emerging establishment, the British East India Company dealt in various commodities including (according to Wikipedia) "drug trafficking, the slave trade, exploration, private military" (sound familiar?) and basically ran India and neighbouring regions firstly for the Company itself and then for the British Crown until the establishment of the British Raj in 1858, just after the Sepoy Mutiny of 1857 where many of the 60,000+ native troops employed by the East India Company turned on their British Officers and sparked a general uprising in India. Through various Acts of government the British Crown then dissolved the East India Company and absorbed it into the British Raj – or direct rule. Thus we see an early model of a commercial operation, infused with selected elites, literally seizing control of half of the world's trade whilst married to the political-military establishment – complete with all of the oppressions and exploitations and devilment that accompanies such 'enterprises'.

In those days of course, and absent our modern communications and ease of travel 'the establishment' was largely confined to discrete geographical regions, and to individual countries or States – or at least to those subject States that they had direct political and military control over. But not anymore. With the rise of new multinational corporations particularly since the 1960's came a concentration of financial pressures on various otherwise-not-connected territories with the singular objective of enriching the corporation. This is how globo-corporate 'values' became engrafted and imposed on national policies, largely through bribery and corruption by vested interests, in this new form of insidious-and-secretive, despotic, international corporate colonialism.

Unbeknownst to the average taxpayer, the resources of the State (natural resources, property and its people) were literally being sold out from under them to the highest bidder, and laws were being enacted by the politicals that no longer served the genuine interests of the people, but instead the financial interests of these corporate-political cabals. Enforcing exploitative laws with private armies of mercenaries still remains on the table of course, but bloody revolutions such as the 1857 Sepoy Rebellion had proven very bad for business. The establishment had learned an important lesson. No..., overt oppression was not being well received by the peasantry and a non-compliant populace was harder and more costly to manage, so 'enforcement' in the modern era would have to be achieved by other more subtle means; by the very nature and structure of the establishment itself perhaps, complete with all of the *appearances* of 'good governance' and of the Rule of Law? In other words, tyranny by stealth and deception with the establishment creating elaborate hierarchies of systemic obfuscation, cunningly masked by convoluted bureaucracies designed to misdirect and mislead the ignorant and the uninformed into believing that the tyranny-by-stealth that is being so effectively visited upon them is actually true 'democracy-in-action'.

In fact, what has actually occurred (although no-one will officially name it as such) is a clandestine return to Middle Ages Feudalism – only this time in an international-and-interconnected, Cabal-organised format. But this is a more perverse form of feudalism where Monarchies have been replaced by the Central Banks; the aristocracy by the super-wealthy and Big Tech; and the Church and its ministers being replaced (or usurped at the very least) by the new 'religion' of money with its self-appointed High Priests, the Bankers, the CEO's of multinational corporations and international moneylenders controlling the cash – arguably the new 'spiritual' (or truly essential) currency in a world dominated by greed and exploitation. And, whereas Middle Ages officials and their agents unashamedly served the sovereign, most modern-day politicians *pretend* to be serving the people whilst actually receiving their instructions from the Banks and global elites. And therein lies another fragment of the Great Lie. Then, we have that great swathe of knaves who have sold their proverbial souls for their own personal piece of the poisonous, State-sponsored pie; the various senior officials and office holders including the judiciary and the Courts. And finally, we have the dupes, fools and moral cowards (the peasantry) who always seem to be the ones who pay for it all.

Hence 'the establishment' today (sometimes known pejoratively as the Cabal) has become a truly global phenomenon with its insidious tentacles in every important pie. It is in effect an undeclared, self-appointed global authority, comprising a great many interconnected individuals – sociopathic by nature or conditioning – whose position in the Cabal depends almost exclusively upon their personal capacity, willingness and ability to generate more wealth, power and authority for the establishment. Achieved through the collaborative use of sociopathic principles and methods of operation, this needs to be squarely understood in context of the Covid-19 phenomenon. Because, as we compare the true facts and emerging evidence against what we are being told by the establishment and by the authorities, one core dynamic appears again and again; that of the truth (i.e. self-evident, verifiable and provable truths) coming smack-up-against the insidious ambitions and multifaceted deceptions of the Cabal. It is all based on lies you see, something we *absolutely* must understand!

There are various theories of who exactly 'the establishment' or 'the Cabal' may be in some of the darker Covid-19 scenarios circulating online, but whether that be "Secret Societies" such as the Committee of 300; Satanists; the Illuminati; the Deep State; Freemasons; elitist paedophile rings, or alien visitors; or more visible targets such as Big Tech; super-rich individuals; the pharmaceutical industry; national governments; the UN; international bankers; the Bilderberg Group; the Davos clique etc., etc., (please do your own research) it is far less important to know *who* these individual players are, than it is to understand *what* actually makes them tick, and how to run down that proverbial clock.

Because as long as we (ordinary people) are willing to give away our sovereignty and sell our souls, there will always be sociopathic predators and parasites ready to step in. And we should be in no doubt that many of the individuals who populate these super-powerful and super-rich cabals actually believe they have a right to assume the role of world leaders – whether elected or not – simply because they are possessed of that singular elitist arrogance and sense of entitlement that arises when you combine the sociopathic mindset with seemingly-unlimited power and money.[vi]

What we should clearly understand above all else however is that many of these individuals have NO moral concept of 'good vs evil' and therefore do not see anything 'wrong' *per se* in their Machiavellian plans. They only think in terms of profits, of self-ambitions, and of how to get ever more power, wealth and influence for themselves regardless of the cost to others. If this requires that they fire a few workers or shut down businesses or even dismantle whole industries to simply acquire profits, then so be it. If more profit, power or prestige is to be had by plunging countries into war, or exterminating whole swathes of humanity to achieve the end goal, well, that remains on the table too of course. Plans are then set in motion to achieve these objectives without any hint of a troubled conscience. That's why it is relatively futile and pointless appealing to their humanity or to their consciences – because in a word, they don't have one. That was part of the original deal with the devil you see. Worldly 'success' in return for your soul, whereby any connection with an original, moral Divine (or soul) is exchanged for the tricks of the trade.

They do understand the concept of 'evil' in an academic sense of course, and they also understand that the bulk of humanity – foolish and naïve as they are – do NOT want evil ambitions to succeed, and so, those nefarious plans must be crafted and disguised and camouflaged as 'social necessities' – yet another part of the Great

Lie. What we really need to be thinking about, is how to disempower these moral deviants, and how to truly empower the good ones amongst us, and that process begins by properly understanding the nature, and the structure of 'the beast'. There was good reason why Jesus said, *"My kingdom is not of this world."* There is also a reason why the Lord's Prayer has the phrase, *"Thy kingdom come. Thy will be done.."* (where?) *"On the Earth.."* !?

Now, without getting distracted by the presence of religious quotes in what is otherwise a Covid-19-related fact-finding exercise; but at the same time acknowledging that scripture often contains hidden truths; the implication here is clear – that whatever kingdoms or empires existed at the time of Christ, they certainly didn't resonate with Jesus as being either 'good' or indeed 'heavenly' entities. One wonders indeed, if there is some link to the history of the ancient Israelites here, and the 'fact' (as the Bible reports it) that God, speaking through the prophets, was very reluctant (to say the least) to allow the Israelites to appoint a king. Up to that time there were only prophet-judges like Eli and Samuel, truly spiritual types with a direct line to the Big Man Himself. But the Israelites were adamant. They wanted a king just like all of the other nations around – nations that God had declared 'unclean'. God finally relented however, and chose Saul from amongst the lowest tribe of Israel – probably in the hope that Saul wouldn't get too far ahead of himself. But we all know how that story ended! After three years of exemplary behaviour, Saul just couldn't help himself. Corrupted by the trappings of power, he became the very model of tyranny itself. One wonders if there's more than meets the eye here?

As a form of sovereignty therefore, God obviously wasn't too keen on kings and emperors, but was more inclined to use 'spiritual' types that He could talk to directly who were grounded in an awareness of Truth. It will later become clear how tyrannies of any sort regardless of their prevalence in human history or how well they are disguised are utter perversions of the notion of one's own 'sovereign authority'. And yes, this includes so-called 'modern democracies' that fall adrift or thrive on endemic deceptions. The historical manifestation of the worst aspects of masculine power and authority such as greed, exploitation, deception, domination, and other injustices which invariably lead to social oppression, poverty, violence and wars; that all of these manifestations of evil can only exist with the tacit compliance and complicity of us all.

So, are we knaves or dupes, or just moral cowards and ignoramuses? Do we want to remain complicit in the Great Lie and all of its eternal consequences, or, are we really ready yet to get 'woke'?

*   *   *

SHAME

My name is shame, and I take the blame
For all that's bad in history
'Cause my friend is greed, and we will succeed
In our destructive legacy

For my tool is man, he who will and can
Pervert the course of history
Because he never sees, what his evil breeds
As he evaluates his treasury.

You see, it's fine to gobble time
And use up precious energy
As long as Mother Earth can take
This rape of her facilities

But what is wealth, without good health?
Or freedom without space?
These very things you're yearning for will leave a bitter taste.

For although I've known you many years
You haven't seen me yet
But you'll taste the shame as we share the blame
For these things you'll yet regret.

CHAPTER FOUR

BEAUTY & THE BEAST

UNDERSTANDING OURSELVES – UNDERSTANDING THE CABAL

Introspection is rarely a comfortable experience, especially when we pose truly searching questions about our own lack of awareness – or of our prior ignorance of crucial things. Harder still if that ignorance has constituted an unwitting complicity in causing others harm. But be assured, if these issues cause you genuine heartfelt concerns, then fear not, because you are already on the right side of the divide.

The concept of 'spirit, mind and body' is central to any understanding of what being 'truly human' means. It is also key to understanding any aberrations of the same. This is a complex multi-faceted topic that we need not cover in detail today in context of the Covid-19 phenomenon, but we do need to grasp a couple of the fundamentals, because they have a hugely important bearing on what's going on. In, *"The Colour of Truth; Amazing Coincidence? ...or Intelligent Design?"* we present an elementary study of 'The Triadic Archetype' which, very simply is the discovery of a undeniable pattern of numbers, symbols, colours and key events in history that cannot be explained other than as interconnected aspects of an archetypal pattern that we are all subconsciously bound to. Not only that, but the same 3-point pattern is evident in atoms and molecules and human DNA, as well as being present in all aspects of nature including light itself. The added fact that all of our wisdom traditions, our folklore and our various religions make specific, but seemingly-oblivious reference to this phenomenon in ways that have not been documented before, raises the possibility that this is a novel discovery for our time; a time when humankind must urgently wake up and understand our own true natures, so that as-and-when we are faced with the proverbial 'Beast of Revelation' that we will not only recognise it, but we will know how to defeat it.

Before we lose those of you who may be atheists or agnostics, let us reassure you that you will not require any particular religious belief to be able to grasp and accept the notion that 'something different' sets us humans apart from the animals. Many will refer to this difference as 'human self-awareness' or 'consciousness' or the ability to reflect upon ourselves; something we are told that animals cannot do. Animals simply act on urge and instinct, and whilst many of them have 'personalities' and individual quirks or habits that set them apart from each other, they cannot be said to be capable of ethical introspection nor of conscious moral decision-making. They do have a brain ('mind') and a body and are therefore capable of processing simple instincts and acting upon them, such as a cow seeking out grass when she's hungry. But making ethical or moral decisions, or reflecting upon the meaning of life? No, that is exclusive to us, and this is where the concept of 'spirit' comes in – in alignment with the fact that we are not mere animals who respond to physical urges such as hunger or thirst so as to find the nearest McDonalds; but that we are also capable of suppressing our physical and emotional urges for a higher purpose – such as considering the health of the planet before buying another Triple Big Mac for instance? We are not just mind-and-body machines in other words (like any beast of the field) but comprise a third element – the spirit, or soul, or conscience if you like, which in ideal 'true human' circumstances would then inform the mind to direct the body. In this manner, 'spirit, mind and body' equates to 'concept, plan and action', and it should *always* follow that order. If we think about this, it makes perfect sense. Other than when falling around drunk, acting in a rage or doing stupid things without thinking (which rarely ends well) everything *productive* in human society follows the pattern; (i) first the concept or idea; (ii) then plan what you are going to do; and (iii) then act upon that plan, making adjustments of course if you need to. But even those adjustments will follow the same pattern of 'concept, plan and action' – right? This 3-point formula is essential to all rational, constructive human activity.

So, if alignment with this Triadic Archetype concept of, 'spirit, mind and body' or 'concept, plan and action' or even 'dream, think and achieve' is the means by which we see 'balance' in all things, particularly when assessing constructive human behaviour; then whenever we are in misalignment with that process, or perhaps when one of those three aspects is either missing or corrupted or out-of-sequence, then we should see a parallel imbalance or discrepancy in the outcomes, which by definition would not be 'good' results – and therefore not 'healthy, happy and holy' (body, mind and spirit) right?

So, if the previous example of acting in a rage or when drunk is equivalent to 'not engaging the brain' (the mind) before getting oneself into all sorts of physical difficulties (the body) – which is why we don't let engineers or surgeons drink on the job – then how much more serious is it if we are all going about our daily lives, raising children and having careers, and building institutions and even whole societies without ever properly engaging the soul? What if everything we know as 'history' thus far is just the record of humankind's failed efforts to create a modern world without ever *properly* engaging the collective spirit or conscience? What if this whole Covid-19 phenomenon – along with so many other aspects of daily life which we haven't properly understood before – is just another disastrous symptom of our collective misunderstanding of the true meaning and purpose of life and of 'the nature of the beast' – to coin a phrase?

The Human Beast: We are going to take a big leap here and drop straight into the deep end with an observation that will come fully into focus when we explore some of the specific why's and the who's of the Covid-19 phenomenon later on. So, if we accept that a 'true' human being is comprised of three parts that make up the whole, and if we call that whole "100%" – then what percentage value should we assign to each of the three parts? The obvious answer is 33.3% each. If we then take a human person who is absent their soul, we are left with an entity that comprises only a mind and a body, or just 66.6% of the whole. Effectively no different to a beast of the field – right? A highly-intelligent beast, perhaps, but still an entity that is absent a healthily-functioning conscience or one might say, lacking a human 'soul'? If we then extend this formula to human agencies, institutions or governments for example that have been created in misalignment with The Triadic Archetype by persons who lack a true conscience, we begin to see the existence of the scriptural 'Beast of Revelation' in existence all around us complete with 'the mark' of that Beast (666) that has been prophesised to take over and destroy the world in a not-very-nice way. Not only that, but inasmuch as each of us is absent that vital spirituality, that sense of morality and conscience that should (in ideal circumstances) be informing our minds and directing our base physical impulses, then perhaps we too are unwitting extensions of The Beast, inasmuch as we carry within us the '666' characteristics of creatures without active spirits or souls? And no, we're not implying that one has to be 'religious' *per se* in order to have a spirit, soul or conscience. After all, God / Allah / Jehovah / Yahweh / Truth / The Divine / The Great Unknown or whatever we want to call 'the source of pure spirit' is not, and cannot be 'religious' in itself. Yes, 'IT' is the object of, and purpose for, religion, but only because we are separated from 'IT' and therefore require some process (religion) to attempt to reconnect with that source. In other words, religion is only a means to an end, and when 'we' (individually or collectively) have achieved that end (full reintegration of the spirit) then religion or religiosity becomes obsolete. (Please think about it).

When we add the poignant fact that The Triadic Archetype (in theological form) is represented by the triad of; (i) Deity, (ii) Wisdom, and (iii) Word-in Action which in turn corresponds directly to 'God, Sophos and Logos' and 'Jehovah, Eve and Adam' and 'Eden, the Tree of Knowledge and the Tree of Life' and/or 'Divine, Feminine and Masculine' respectively (the theologians and religionists should understand) along with a great host of accompanying symbolism all the way from the story of the mythical Garden of Eden to current events occurring in our modern worlds, the quote from Revelation 13:18 suddenly takes on a different and possibly very important meaning. Making allowances for the original Greek and for scores of variations of the exact, precise translation (which is literally, not possible from ancient Greek to modern English) the verse says:

"Here is wisdom. Let he who has understanding calculate the number of the Beast for it is the number of a man (or mankind). That number is 666."

Selected aspects of The Triadic Archetype

Spirit	Mind	Body
Concept	Plan	Action
Dream	Think	Attain
Believe	Conceive	Achieve
Morality	Ethics	Pragmatics
God	Eve	Adam
Deity	Wisdom	Word
Divinity	Sophos	Logos
Divine	Feminine	Masculine
Eden	Tree of Knowledge	Tree of Life
Soul	Conscience	Integrity
Love	Compassion	Strength
Eternal	Internal	External
Holy	Happy	Healthy
Choice	Environment	Genetics
Electron	Neutron	Proton
Air	Water	Fire
Nirvana	Yin	Yang
Art	Philosophy	Mechanics
Truth	Thought	Reality
Ideology	Theory	Practice
Universalism	Collectivism	Individualism
33.3%	33.3%	33.3%

And these are only a few examples that are relevant to our topic today. The main concept to grasp being the possibility that the original model of spirit-to-mind-to-body has been corrupted first of all; (i) by the separation from the original (divine) spirit, and secondly, (ii) by the reversal of the feminine-to-masculine dynamic, whereby the worst traits of the masculine have dominated human history. But without straying into convoluted hypotheses about who or what this may all mean in context of the Covid-19 phenomenon (more on this later) and not wanting to depart from the factual, evidence-based approach that we hope to maintain as we tackle a number of contentious Covid-related issues, let us simply leave the question open as to whether or not it is possible – or indeed probable – that human history to-date has been the story of the unthinking and often spiritually-corrupted actions of the unrestrained masculine 'Beast' a creature comprising only a body and a brain (66.6% of the whole) who is devoid of Divine guidance (spirit) and largely absent the feminine maternal-and-empathetic influences and conscience-based wisdom that would flow from such a spiritually-enlightened mind? That base creature, the subject of myths, legends and historical reports has fated humankind to an endless cycle of violence, of viciousness and greed largely dependent upon the organisations and institutions that are spawned by those very vices. A cycle of institutionalised, collective-and-compliant conformity which will only ever be undone when we as individuals awake from our collective apathy and ignorance, and re-imagine our lives in context of The Triadic Archetype.

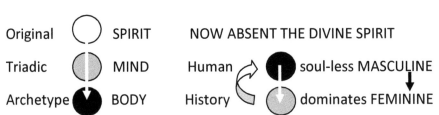

Original ⦿ SPIRIT NOW ABSENT THE DIVINE SPIRIT

Triadic ◑ MIND Human → soul-less MASCULINE

Archetype ● BODY History dominates FEMININE

There is a reason after all that it is called 'HIS'-story. Perhaps it now needs to become OUR story – a story that incorporates not only the ultra-dominant masculine elements in their *proper* role of strength-

and-service, but the guiding (holy) spirit and aligned feminine wisdom as well? (For more, please read the books).[2]

Briefly, on the importance of symbolism and why it matters. Now this may take a few runs at it before the concept fully sinks in, but in context of the Covid-19 phenomenon, and indeed in context of any complete understanding of the proverbial 'meaning of life' it is essential that we understand the place and role of symbolism in human history. This has an important, direct bearing on our Covid-19 discussion today. The concept is simply this: we usually think in terms of two forms of 'reality'. First of all, there are our personal beliefs. Then, there are facts-and-things that we can see, feel and touch. Inasmuch as I can *believe* anything I wish until such time as the empirical facts prove otherwise, then clearly and logically, tangible 'facts-and-things' take priority in any manifest world over mere imaginings or 'beliefs' – agreed?

Okay, and with respect to those amongst us who may prioritise their religious *beliefs* over the events of daily life, or, who accept and 'believe' the official narrative without seeing or knowing the proofs; we now need to introduce the concept of two other levels of 'reality', and it is here that we re-enter the world of otherwise inexplicable archetypes – those subliminal phenomena that affect us all whether we are consciously aware of it or not. These other levels of 'reality' are, (a) symbolism, and (b) 'perfection'. Understanding that we are merely skimming over a very profound set of concepts here, but equally, in the understanding of how very important it is for us NOT to dismiss specific symbolism (usually presenting in the form of certain recurring archetypal patterns, colours and numbers) as being some form of arbitrary happenstance that may, somehow, be interestingly coincidental to this-or-that, or which adds romantic colour or flavour to ancient myths and legends, or which appears in arcane quotes in our various scriptures for example; let us for now simply consider the possibility that humanity is NOT living up to its

---

[2] (i) "Psychology, Symbolism & The Sacred" (ii) "The Colour of Truth" (by this author)

potential as a global society populated by individuals who are each comprised of a healthy body, a happy-and-wise mind, and a holy (Divine) spirit working in perfect and fruitful collaboration in good and productive societies.

If we then consider the possibility that there is in fact an original plan or blueprint – a universal, archetypal plan – that points to what such a 'perfect' human society would look like, but, because it has actually never existed in historical format in human civilization (other than theoretically-theologically in the mythical Garden of Eden) then the only way to make necessarily-indirect reference to any such state of 'perfection' is by way of subliminal signposts if you like, in the form of certain patterns, symbols, colours and numbers that exist all around us. But being symbols, they can only *point* to that original ideal, vs being the ideal itself. Thus, archetypal symbols that appear in so many mythologies such as the Tree of Life (masculine) and the Tree of Knowledge (feminine) which in turn align with key properties of the electromagnetic spectrum, suddenly take on a new meaning.

Spirit	$(\infty)^3$	White	God	Inspiration	Concept
Mind	40 or 400	Blue	Eve	Knowledge	Plan
Body	70 or 700	Red	Adam	Life	Action
Evil	**6 or 666**	**Black**	**Satan**	**Death**	**Destruction**

Similarly however, there are symbols, colours and numbers that serve as a stark warning of approaching evils, which we ignore or dismiss at our peril. Because 'the opposition' who deal in these evils are very much aware of the importance of such symbolism, the most current of which arguably, is the proverbial 'mark of the Beast'. Given that *The Colour of Truth* book will answer many of those

_____

[3] The symbol of infinity (a sideways '8') is here encased in the eternal circle.

questions, let us simply summarise the concept like this. The colours and numbers in the table come from the corresponding physical properties of light itself where deep-reds and deep-blues mark the limits of our visual perceptions comprising wavelengths measured in nanometres.

Could it be mere coincidence for example – at odds of billions to one – that before any of them could possibly have been consciously aware of the scientific properties of atoms and molecules, or of the chromo-numeric properties of light itself, that so many of our ancestors and sages and prophets – not to mention kings and tyrants and emperors and the societies they ruled over, made continual, subliminal, but nevertheless *direct* reference to the properties of The Triadic Archetype?

This forces us to ask the hugely pregnant question of what exactly is the connection between these perennial symbols and our daily reality? How do we interpret and apply this symbolism to human society and to each of our lives in such a manner as to align with the original model of 'perfection' especially in context of what's happening today? Because if we can align ourselves as individuals and as societies with the properties of The Triadic Archetype then maybe we might yet still have a chance against the wickedness of man? You see, although we are living in the midst of a natural world and a cosmic universe that aligns exactly with the patterns and principles of The Triadic Archetype; human society does not! Far from it! Absent a direct connection with the infinite Great Spirit through the suppressed feminine, we are doomed to repeat the red-black absences of wisdom and of truth, and of moral inspiration, thereby creating a veritable 'kingdom of beasts' populated by soul-less knaves and dupes which is dominated by the violent excesses of the morally-indigent masculine. An intrinsically evil masculine that will always resort to violence and wickedness to defend his position as a lord of destruction and exploitation in the proverbial 'kingdom of hell'.

The comparison is obvious. 'Perfection' would equate to the Kingdom of Heaven, whilst a world that is absent the guiding spirit of truth and goodness can be called nothing other than the Kingdom of Hell. So, if the Kingdom of Heaven that Christ spoke about 2,000 years ago was not, "on this Earth" at that time – and if nothing much has changed since in respect of the destructive behaviours of men – then maybe we are all still living in the Kingdom of Hell – on Earth? A world governed by lies. Because undoubtedly, all of the worst traits of the corrupted masculine are still dominant in our ruling classes and in the various institutions of the State, with calculated, systemic deceptions at the heart of it all so as to mislead and exploit the public. Lies indeed seem to be the currency of the day – most especially those of the Cabal. And who was it that was known as "the father of lies"?

In this manner, scriptural terminologies that encapsulate the notions of 'good vs evil' and 'truth vs lies' or even that of 'the Beast' as individual persons or as humanity as a collective can be accurately applied against; (i) the goals and objectives; (ii) the motives-and-methods; and (iii) the structures and institutions of the modern world, and this is where the term 'diabolical' truly comes in to its own. You see, despite romantic notions about personal freedoms and our perceived independence within functioning democracies, it appears that we are all in effect, just unwitting cogs (at best) in a malevolent Machine that relies largely on our ignorance and our apathy to press ahead with 'the agenda'.[4] As with any covert enterprise, some are more aware than others, and some are more complicit. The agenda may take many forms in different spheres and locations, but it is *always* predicated on deceptions and it *always* rewards the deceivers – or at least, that's *always* the objective. The bigger and bolder the agenda – the more knaves you need to push it and the more compliant dupes you need to swallow it, to comply with it and to pay for it, and this unfolding Covid-19 phenomenon is no exception.

---

[4] See, "Humantruth: A Philosophy for a World in Crisis." by John Bapty Oates.

CHAPTER FIVE

SOCIOPATHS vs EMPATHS

THE DEVIL IS IN THE DETAILS

If 'the agenda' (whatever it might be) is a deception, or, if it is predicated on lies, then control of the official narrative becomes essential to those driving that agenda because obviously 'the official story' has to be controlled by liars, whilst the real truth must be simultaneously suppressed. To achieve this level of coordinated deception one needs a well-trained cohort of practiced deceivers, amoral and unashamed, who see profit in all manner of deviousness. In other words, one needs professional liars and similar others who harbour no moral scruples to sell the script. And this is how and why, over a period of several decades, we have arrived at a place where the highest percentage of sociopaths and psychopaths (by profession) who are undoubtedly 'knavish' by inclination, populate the Executive Offices of multinational corporations, with the legal profession coming a close second, and 'the media' being listed right behind.[vii]

Well, why should that concern us, you might ask? After all, most of us will never have cause or opportunity to grace those affluent locales. Well, very basically, it should concern us because according to the psychology; sociopaths and psychopaths lack certain 'humane' traits such as compassion, empathy and remorse. One could say that they are absent what others would call 'a conscience' which makes it easy for sociopaths to deceive, exploit and abuse others as they ruthlessly pursue their own narcissistic goals. This almost invariably includes the pursuit of power and money. For a full explanation of the sociopathic condition, please read, *"The Rise of the Sociopath – and why YOU should be worried!"* because this troubling phenomenon lies at the heart of all of the trauma that we are facing today – both at the individual and social levels – as well as in corporate-political spheres.

The Empathic - Sociopathic Scale
Fig 1. The Theoretical Model

Excluding any other factors, it is ONLY the 'top' 5% of the population who have influenced ALL of the change in history to date.

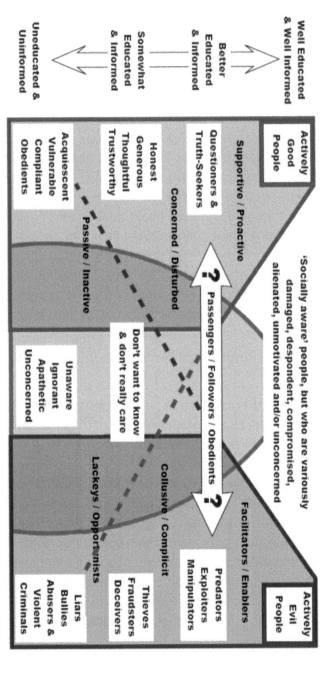

'Socially aware' people, but who are variously damaged, despondent, compromised, alienated, unmotivated and/or unconcerned

Well Educated & Well Informed

Better Educated & Informed

Somewhat Educated & Informed

Uneducated & Uninformed

Empaths

Apaths

Sociopaths

Actively Good People

Supportive / Proactive

Questioners & Truth-Seekers

Concerned / Disturbed

Honest Generous Thoughtful Trustworthy

Acquiescent Vulnerable Compliant Obedients

Passive / Inactive

Passengers / Followers / Obedients

Don't want to Know & don't really care

Unaware Ignorant Apathetic Unconcerned

Collusive / Complicit

Lackeys / Opportunists

Facilitators / Enablers

Predators Exploiters Manipulators

Thieves Fraudsters Deceivers

Liars Bullies Abusers & Violent Criminals

Actively Evil People

But for clarity here; we equate the notion of 'sociopathy' as being akin to operating without a soul, a spirit or conscience, and absent the feminine and empathetic traits of compassion and (true) wisdom. 'Beasts' in other words, who are not attuned to any moral compass.

To avoid unnecessary confusion on a topic that may be alien to many of us; although both psychopaths and sociopaths share many of the same personality traits, for the purposes of this book the main difference to note is that psychopaths are either born that way or have suffered brain damage (for example) that renders them literally *incapable* of empathy. Whilst sociopaths on the other hand can be trained, bullied or abused into the condition, either through trauma or social conditioning – or indeed as a requirement of their jobs. Think of the ruthlessness of violent criminals for example, or the callousness of CEO's who must pursue profit any price, or, of the necessary absence of empathy – even temporarily – when a soldier must kill. In fact, any social circumstance or any occupation that requires the ability to objectify the other may be instrumental in producing sociopaths who range across a broad scale of sociopathic severity, and who arguably therefore – in contrast to the irredeemable psychopath – can possibly be reformed or 'redeemed', so-to-speak.

Again, to explain very briefly; at the opposite end of the scale are the genuine empaths; the caregivers, nurses and teachers who share traits like compassion and understanding and who genuinely 'feel' for others, especially when they are in need. And somewhere in the middle is that great mass of humanity comprising the apaths – the apathetic amongst us who, for various reasons to do with education, social conditioning and personal experiences lack the knowledge, the will or the resolve to actively participate in social change, either constructively or destructively. In Boétie's study, these apaths would be the dupes. But both active empaths and inactive apaths – whether or not they are aware of it – are viewed variously by the psychopath (knaves) as either fools or as cannon-fodder; as products or prey; or as mere objects or resources to be used, abused, manipulated, exploited, discarded, ignored or destroyed as needs be.

The main point to note here – and it is an extremely important concept to grasp as we move forwards in context of the Covid-19 phenomenon – is that certain types of people are drawn to certain occupations by way of the fact that their very personalities predispose them to engage in routine deceptions, exploitations, manipulations or even the casual destruction of others, as they pursue personal power, prestige and possessions (wealth) with callous and ruthless ambition. In other words, they are perfectly suited to a capitalistic environment that requires, as a basic principle, that you put profit before people, and, each time that you do so, you will be rewarded accordingly. Not very nice people then, one would imagine!?

But just hold on a moment! Because the opposite is actually often the case. Because if they need to be 'nice' or charming to get what they want, then they will play nice very convincingly – with the emphasis on 'play' because ultimately life is just a game to sociopaths, and we (the rest of us) are their playthings. As long as we keep entertaining them, or are making money for them, or are having sex with them – or whatever it is they need from us, then you can expect to be charmed or groomed as required by an individual to whom lying and deceiving is second nature. Or perhaps more accurately, their first nature? Don't forget that these are people to whom conscience is an irritation. There is no genuine 'spirit' *per se* behind their decisions, other than a spirit of deceit, and there is no genuine compassion or concern for those they exploit. As long as the energy is flowing their way, you may get no indicators at all that you are in effect, no more than an object to them. Something to be used and exploited without conscience, regret or remorse.

Consider the many lies and false promises being made so convincingly in election campaigns, and how casually afterwards those politicians can 'carry on regardless'? The deftness with which they evade pointed questions, and their capacity to deceive without remorse or regret is one of the first give-aways that you are listening to a sociopath. Indeed, being masters of the lie whilst having the ability to project sincerity on camera are two of the fundamentals for a career in public office. Please, do NOT expect the truth. Do NOT expect honest introspection, and do NOT expect any sincere apologies when they get caught out. They will of course 'say sorry' or make an outer show of concern if that is required to advance their agenda, but it won't be real. It rarely is. These are just means to an end, and that 'end' will almost always be at our collective expense.

If the sociopath can get what they want through other means, or if those other means are more practical or effective to their goals for example, such as using bullying or threats or violence, then that's what they'll do, and again, without any remorse or sense of conscience. And as-and-when they finally tire of you, or if you stop performing or when they simply 'find a better option' they can discard you like a used rag without any sense of regret or responsibility. In fact, sociopaths gain a perverse pleasure and sense of empowerment from hurting others especially if they can do so with impunity because they get no real joy from being 'humane' in the classic sense. The usual human emotions that make life so fulfilling for the rest of us such as love and kindness, and true, deep friendships mean little to the sociopath, unless the *pretence* of the same is in their interests. And whilst these type of people may be very efficient in making money or in running concentration camps for example, surely the question needs to be asked, are they really suitable to be running the show? More poignantly perhaps, maybe we should be ruminating on the dynamics that must be in play in those settings where sociopaths and psychopaths can actually thrive? Why is it for example, that we didn't see this problem before? How and why is it becoming such an issue today, and how did so many of these 'knaves' get to the top?

Well, the simple answer is that this is the creeping result of the proliferation of capitalistic 'values' along with decades of policy decisions that have favoured the establishment over the general common good. "Profit at any price" in other words, and thus the proverbial scum has risen to the top. This has come on the back of centuries of social conditioning by the powerful to convince us (the majority) of the absolute and largely-unquestioned 'need' for top-down governance. Previously, there were probably just as many potential sociopaths out there, but the conditions weren't nearly so favourable. Now you can be a sociopath by profession – and get very well paid for it. The result is that 'they' are now in charge, and we, as individuals who might aspire to something better, are effectively disempowered. In context of the unfolding Covid-19 phenomenon, it seems that we must now accept what we're told and then DO as we're told, or we WILL suffer the consequences.

Sadly, most of us have been too busy 'earning a living' or have been too preoccupied with mind-numbing distractions and debilitating, media-projected fears since the start of the Covid crisis in particular, to actually realise what has been going on right beneath our noses. Others can see – or at least they suspect what's happening – but are unable to process it rationally. Some call this type of angst-based denial 'cognitive dissonance', which is a fundamental unwillingness or inability to accept new ideas or notions that conflict with our often-unrecognised social conditioning, or with our pre-existing beliefs. For example, why would the Chinese authorities' first reaction to news of a viral outbreak, be to try to cover it up? Because even the dumbest Party Official must know that viruses are not subject to the Official Secrets Act, nor do they respect curfews or obey orders – even from hard line communists.

Well, there are a number of possibilities here, some of which are more disturbing than others, especially in light of the UK Government's official declaration of March 19th 2020 that stated:

> "As of 19 March 2020, COVID-19 is no longer considered to be a high consequence infectious disease (HCID) in the UK."

Whaaat!? NOT a high consequence infectious disease?? Well, what the ruddy hell is it then – and what's all the panic about – right? Any person with half a brain must be seriously taken aback at a statement like this given all the subsequent social upheaval. But at the time, the statement was patently accurate, because the UK didn't register its first reported death from Covid-19 until March 5th 2020.[viii] Yup, you heard that right as well. Even with all of the outstanding questions about what 'Covid-19' really was/is, even the UK government acknowledges that no-one had died from "it" until March 5th 2020.

Now we need to make a very serious point here. Because many people will have just read that statement by the UK authorities and will simply not be able to process it. It is such a glaring indictment of the whole shut-down and lock-down reaction as to beggar belief. Because a '<u>low</u>-consequence infectious disease' by definition, would be something like the common cold or flu, and if that's the case, then what's the whole Covid-19 global health crisis all about, then?

Remembering that we are now going back to mid-November 2019 to February 2020, we need to be asking what it was about Covid-19 that had the Chinese authorities so worried (apparently) because clearly it wasn't the 'high consequence infectiousness' of Covid-19? Or, maybe it was? Because arguably, they didn't really know what they were dealing with at the time, did they? Surely we all remember the videos of rows of dead bodies on the sidewalks, waiting apparently for collection by the morgue, and of ranks of fumigator trucks spraying the streets through the night with bleach and disinfectant as well as the 76-day total lockdown of Wuhan that affected 40 million people? What was all that about if this wasn't a 'high consequence disease'?

Trusting What We See: Well, one of our first points of clarity is that the videos of 'dead bodies' was NOT shot in Wuhan, and those were NOT dead bodies. They were just people sleeping rough because of the Covid restrictions that the local government had imposed in Shenzhen.[ix] But by the time the video clip was spotted as a fake, it had already been seen by millions, and the damage had been done. That is, that great alarm had been spread about the high mortality rates of this new viral plague. The Twitter account that posted that clip is now gone, but the original poster's account *looked like* that of an American patriot.[x] But as we said earlier, with so many lies and misinformation out there, and with so very much at stake, every piece of information – from social media to the MSM – must be checked *objectively* (i.e. without subjective prejudices or foregone conclusions) before coming to firm decisions about the truth of any given claim. Whilst we realise that this is an overwhelming task for most of us, and that it is far easier to just accept the official narrative. Nevertheless, we have to remain alert to the fact that the establishment is absolutely relying on the credulity of the masses (as they have done for centuries) to swallow whatever it is they are selling us, so as to keep them in power and further their agendas – whether personal, political and/or financial.

Secondly, yes, there are some people out there who get their thrills seeing *their* fake news being taken up en masse. So, whilst the overall message of this book is that 'the authorities' (in general) have demonstrated time and time again that they simply cannot be trusted, that we also need to be alert that those who criticise the system are not necessarily rational, balanced and credible either. We need to be discerning in other words, and commit to the truth.

Interestingly, one of the sources that exposed the Shenzhen video clip as a fake in February 2020 was a medical student who commented:

> *"The mortality rate of the coronavirus is very low as compared to the SARS (Severe acute respiratory syndrome) which hit China more than 15 years ago."* [xi]

This insightful observation (which will come into play later on) that has since proven to be absolutely accurate was not however seized upon by the MSM or by the authorities as they continued to push the 'dangerous new Covid plague' narrative. So, why ever not? Why such oppressive and draconian measures by the authorities one might ask? After all, if there really is a genuine and immediate threat to the health of humanity, surely all of the sensible adults out there will acknowledge and accept this fact – based on the evidence and on expert opinions – and will surely 'do the right thing' for the sake of all; and will do so voluntarily, upon their own recognisance? After all, even during catastrophic natural events such as severe storms that seriously endanger the population, the authorities have rarely issued such oppressive and restrictive orders, relying instead on the common sense of grown adults to 'take the appropriate measures' to protect themselves – just like we do every flu season for example, with the old and the vulnerable taking sensible precautions in face of a disease that could, possibly kill them. Think isolation, chicken soup and possibly even the latest flu vaccine?

But apparently, this new virus is different. This one requires that we are all locked down, isolated, and masked-tested-and-vaccinated, and it has to be done RIGHT NOW (apparently). If we do not comply with these diktats, then woe betide us, because we will (as we have seen) be variously bullied, coerced, tricked, shamed, threatened and intimidated into doing so – and then punished if we do not! But most people don't like being punished, especially when they don't really understand why. No great surprise then, that disturbing questions are now being raised as to the logic and efficacy of lockdowns and plenty of *dissenting* expert opinions are emerging on what is actually going on and what the Covid-19 phenomenon really is? Somewhat disturbingly, it is largely those who are *not* embedded in the establishment and who regularly seek out the truth and the facts independently for themselves who are becoming the loudest voices in opposition to the official narrative.

"What! Are you saying that the authorities can't be trusted? That they have been lying to us all this time? Because if that's the case, then I am a dupe, a fool and a patsy, and lots of people that I know and love are exactly the same! And that would also mean that the media and the authorities are taking us all for fools. No,

that just can't be true, and I simply won't have it! So, you can consider yourself blocked!"

This type of ignorance-apathy-and-fear-based 'voluntary servitude' however, rooted in elemental denials of fact, is a long-ingrained habit dating back to ancient times. Conditioned from birth so as to fulfil our roles as compliant obedients; as slaves and labourers; as efficient producers and persistent consumers; as reliable taxpayers and good little soldiers; few have had the time or inclination to reflect upon 'the great big lie' and the part that we unconsciously play in its relentless propagation and proliferation. Like an en-masse case of Stockholm syndrome, when kidnap victims and prisoners start to unconsciously identify with their abusers; we have been slowly and systematically lulled into a collective apathy through the patterns, systems and institutions of history. Provided the tyranny wasn't too bad you see, most societies have been happy enough to defer to central authorities and political institutions with the great bulk of the population accepting the need for various rules and regulations and the social requirement that they obey directions and instructions 'from above' – despite the obvious moral flaws and corruptions that accompany that acquiescence – in return for a comfortable and untroubled life that is devoid of major distress or of difficult decision-making. Thus the habit of delegating one's sovereign and moral authority to some centralised power has become 'the norm'.

In this manner, a two-tier society has long been in existence and it has been accepted (albeit unconsciously by most) that this "us-and-them", or "rulers and subjects", or "bosses and employees" dynamic is essential to the functioning of human societies, and that whatever personal compromises must be suffered to keep the Machine ticking over, well, that's just a case of, "the lesser of two evils" – right?

Wrong! It will always be the greater evil to abandon our sovereignty, our personal moral responsibility, and our human dignity and integrity. Even in those cases where we have lauded and applauded

70

great social developments such as the establishment of 'universal education' by way of the creation of schools and colleges for example or 'free health care' for all, there is always a price to pay for divesting our moral responsibility to 'the establishment' and to the hierarchical institutions set up by them, to run things. In the case of education for example, it takes quite a leap of perspective to grasp the possibility and indeed the reality that the reason that education was suddenly made universally available to the peasantry was first of all because there was profit in it for the establishment, and secondly, because the parallel objective was never to fully nor properly educate the masses in that which would truly liberate them, but rather, to indoctrinate and condition them into becoming good little producers, consumers, workers and soldiers for the Cabal. Dupes, slaves and servants in other words who, because they believed themselves to be free, well-informed autonomous beings, would then more willingly surrender to the lies and seductions of the diabolicals.

Going back to one of my own personal sources of inspiration, we ask the fundamental question that should, in any properly-functioning conscience or indeed in any honest society or organisation be at the very heart of it all. Am I doing good or am I doing evil? And how are we supposed to know which is which?

Well, this leads us to the issue of personal morality and the long-held belief that basically, it is up to me to decide what is right-or-wrong in my own life, and that I have the right to do as I please as long as I'm not breaking any secular laws – right? I mean, we all understand that if we get caught speeding or robbing a shop that there will be consequences – but *only* if we get caught – correct? And this is where many of us, as individuals and as societies are missing what is possibly *the* most important point of all; that if my own personal sense of morality – or indeed that of the collective (wherever we happen to live) – does not align with Universal Law, then there will, absolutely be consequences. This is not a case of whether or not we get 'caught' because Universal Law is written into the very fabric of the universe itself, and the consequences of violating it are inevitable and they are absolute. In fact, the whole concept of 'human evil' is precisely that: it is the repeat, continued violation of Universal Law by a largely ignorant and unaware population. But there are those too, who are profiting from our communal ignorance, our collective sins if-you-like, and they know *exactly* what's going on. What's more, they couldn't care less about the concepts of good or evil or of moral rights or wrongs, as long as there's profit and reward in it for them personally.

If we think about this for a moment it should make perfect sense. Most of think that as long as we are not breaking community laws that basically anything goes. But, just as we don't allow kleptomaniacs (thieves), serial murderers or child abusers to have free licence with their 'natural' urges because of the damage that would do to society, so must we start to examine what other types of generally-accepted human behaviours might be violating Universal Law. Just because we have passed laws that allow this-or-that for example, does not necessarily mean that 'it' (whatever 'it' may be) is morally right or wrong, especially if it can be shown to be a violation of Universal Law.

Naturally, I am suggesting that anything that is not in alignment with Universal Law in the form of the Triadic Archetype as presented, provisionally, in this work is arguably a 'moral' breach of Universal Law which will, inevitably, have 'other-than-good' consequences for us. You see, much that is happening right now under the umbrella of the Covid-19 phenomenon and the emergency legislation brought in by the Cabal to endorse and legitimise what the Cabal is foisting upon us; particularly the unnatural, oppressive and indeed soul-destroying restrictions on our very expression as sovereign and spiritual beings, are in-and-of-themselves inhumane in essence, and, despite their purported 'legality' are in effect gross and serious violations of Universal Law. The fact that it isn't *technically* against the law for me to try to fly off a building – but it *is* in violation of the natural laws of gravity – well, that's one good example of where human legislation does NOT necessarily align with natural law. I may have genuinely *believed* that I could fly, but guess what!? Our beliefs, our notions or our politically-correct thinking can be legislated *ad infinitum*, but in the end it all means nothing if it isn't in alignment with Universal Law. In fact, if it isn't in alignment then it is, by definition, an *interference* in that ultimate Law and by further definition comprises an act of human evil. It is not necessary in other words for there to be conscious malintent for something to be essentially evil. Malicious intentions certainly help expose what's wrong, but human evil in so many forms, is largely predicated on ignorance, apathy and the collective abrogation of our moral responsibility to properly understand the Earth that we inhabit and the societies that we live in. Remember the fundamental 'concept, plan and action' theme where; (i) the concept is rooted in Universal Law, (ii) the plan is based upon a well-informed understanding of the mechanics and the parameters at play, and (iii) the resultant action should therefore be 'good' in essence. But remove any of these factors from the equation, and at best, we can expect some sort of messed-up result, which arguably, because it does not properly align with Universal Law, could well be defined as variously, 'wrong, broken, amoral, incorrect, dysfunctional' or indeed 'evil'. Accordingly, the idea that any one of us can simply do as we please based upon our own personal choices, appetites, urgings or beliefs, irrespective of Universal Law has to be completely rethought if we are ever going to solve the problem of human evil.

Reflecting upon the three temptations of Christ in the desert, which, whether you are a religious person or not nevertheless informs us of certain subliminal themes that are common to all scriptures and mythologies; we see some immediate but largely not-spoken-about clues about the nature of evil and the insidious methodology it employs, noting the curious fact that each of those demonic temptations aligns neatly with the notion of The Triadic Archetype.

Satan's first temptation was physical – to do with the body. "Why don't you turn these stones into bread Jesus?" The second temptation appealed to Jesus' vanity (his mind) – of how he *thought* of himself as being so important that if he threw himself from the Temple that angels would surely catch him. And finally, Satan invites Jesus to bow down and worship him from a mountaintop in return for, "..all of the kingdoms of the world!" This was the spiritual temptation, or the temptation to sell one's soul; to surrender one's own divinely-inspired sovereign moral authority – one's connection with the Divine – along with the enlightened consciousness that flows from that and hand it over to the Prince of Lies in return for worldly rewards. The fact that the 2nd temptation was delivered from the pinnacle of the Temple – supposedly the holiest building on Earth at the time – should not be overlooked. The '666' symbolism of a pinnacle-tower that measured 180ft might also have import here.

Thus we see that the journey to truth begins with us understanding not only the importance of the universal 'spirit, mind and body' paradigm in each of our personal lives, but also of that original natural order as modelled in The Triadic Archetype – universal in all of its manifestations other than in corrupted human society – and inclusive of the realisation that 'the Cabal' of today; that pernicious worldly establishment that would control our individual destinies via guile, diabolical temptations, threats and rewards, is rooted in the Great Lie.

This Great Lie – the basis upon which so many human institutions and organisations operate – is a gross and truly diabolical perversion of that original universal concept of personal, and responsible sovereignty. This perversion is not only evident in the psychology of the individual membership of the Cabal, but is also central to its operational structures and ambitions.

CHAPTER SIX

ORGANISED CHAOS

DYSFUNCTION BY DESIGN

All organisations and institutions can generally be divided into either; (a) private initiatives such as businesses or corporations, or (b) public-interest entities such as government and all of its branches, with the single major difference between the two being that private businesses need to make money to survive, while governments can simply take the money they need from the population. In other words, businesses generally need to be efficient and effective in the use of their resources, whilst governments do not. Not as long as a gullible, ignorant and compliant public keeps paying their taxes, fines, and generous salaries for government employees.

In context of the Covid-19 phenomenon (and indeed many other aspects of modern life) it is crucial to understand these fundamentals especially when considering the proposed 'marriage' between the private and the public sector that we seem to be hearing ever-more about from national governments. Because this is a 'marriage' between private entities (businesses and corporations) that are – as a first principle – profit-driven; and between other entities (government departments and agencies) who not only do NOT have to worry about immediate 'profits' *per se* (because they are taxpayer-funded) but who are responsible for wholesale, reckless squandering of public resources and whose very justification for existence may be open to serious question? In other words, that whilst the continued existence of most privately-owned businesses is dependent upon their ability to generate profits in a competitive marketplace, that particular element of 'competition' is notably absent in government agencies. And there, in one short paragraph we have a logical explanation of how-and-why there is so much inexcusable waste, and inefficiency, and unnecessary bureaucratic task-duplication within governments. Because the very existence of some government agencies and departments – and even certain whole governments in-and-of themselves – is predicated NOT upon any essential societal need (although they do like to claim that to be the case) nor upon any requirement to generate profits or return value to the people; but upon that entity's internal need to justify its own existence. In some cases that existence is qualified – albeit usually deceptively – by ALL governments' need to have a stable, loyal and reliable work force (however inefficient) in order to be able to run the show through sheer weight of numbers. This, somewhat euphemistically, we call the civil and public services where apaths and incompetents of various hues combine mindlessly with unscrupulous knaves to keep the cogs of the establishment churning.

Well-salaried positions on public bodies such as Ireland's now world-famous parade of toothless 'oversight bodies' and other self-serving semi-state quangos illustrate perfectly how the establishment further rewards dutiful compliance with 'jobs for the boys'. And this is how the stage is primed for the toxic takeover of public services by profit-driven corporations. Because, if we think about it for a moment from the perspective of a sociopathic CEO of some multinational corporation (such as Big Pharma for instance) what a master stroke of business genius it would be if we could tap into all of that 'free' money that governments have access to – right? Especially in circumstances where the public are so used to getting a poor value service or product from the government that they will hardly notice the difference when previous government waste and inefficiency is transformed into obscene profits for the corporations?

In this manner – and for these express reasons – governments variously 'privatise' or 'nationalise' different aspects of society, or initiate so-called 'government partnerships' involving dodgy deals with private law firms or with big finance for instance, and the big question we (the public) should always be asking, is who is making these decisions, and why are they making them? Because the opportunity for ruthless profiteering – at the expense of the general public and of humane values – is overwhelmingly evident. But whatever their motives and intentions, and whether private enterprise or public institution, all of these entities need a basic operational structure which almost invariably falls into the hierarchical top-down model which is all fine-and-dandy if the person at the top of the hierarchy is a saint. But what then, if they are not?

Human organisations and institutions comprise, basically, the known history of humankind in the form of ancient communities, kingdoms, nations and empires for example, as well as all of the internal operational systems within each entity required for governance such as education, commerce, war, or the management of resources. All of these entities depend upon top-down hierarchical structures (usually pictured as pyramids) that have assumed the authority to do this-or-that in any particular setting. These various institutions and organisations are so pervasive that most of us hardly give them a second thought, because that's simply the way the world works – right? After all, we all understand the economics of good organisation and the practicalities of having institutions that carry out necessary functions on the public's behalf in the genuine interests of all. But there is one massive, fundamental flaw in the hierarchy of any organisation once it grows to a certain size, because no matter how solid the integrity of the leader / boss / president / head of the organisation, he/she can only control the intrinsic values and integrity of the organization to the limits of their own personal influence, and this, I would argue, extends only to about three levels of function.

The concept of 'three levels of function' is not as complicated as it sounds, and will obviously vary with the type and size of any given organisation, but it is simply this; that if you join any organisation, business, sports team, religion or political party (for example) based upon the declared ethos / vision / beliefs of the central figure in that organisation (the boss / the manager / the leader etc), then unless you are within three levels of function of what is being said in the boardroom, at cabinet, or in executive meetings, and unless you can marry your own conscience to what is being planned, that by the time the instructions pass through just three tiers of management, that those instructions will have inevitably become distorted or corrupted by the self-interests or other-than-noble ambitions of someone in the chain of command. This happens whether the organisation is in itself intrinsically wicked or, whether it declares itself to be intrinsically good, which is how and why members of the Mafia for instance, who believe themselves to be true and loyal soldiers to the Boss can go in to a meeting with their head held high, and then come out, feet-first. Or, how an honest and devoted nun can serve the Order diligently all of her life unaware that her Mother Superior is covering up decades of child abuse. And this is how so many otherwise decent people who are just 'earning a living' or just 'doing their jobs' find themselves unwittingly doing the devil's donkey work. Yes, if you join the Mafia then you can probably expect to get blood on your hands, but that's not the point we're making here. The point is that by joining any organisation, or business or group we almost invariably give away our moral authority – or at least some of it – in return for 'instructions from above'. It's a trade-off between the individual and the organisation, with the organisation basically telling you to park your conscience, or morality or intuition in return for a wage. Usually we call it 'a job'. Others might call it something else. The curt rebuttal from the sergeant to the new recruit who 'thought' he was supposed to do this-or-that sums it up neatly: "You're not paid to think, laddie!"

The fact is that most of us tend to seek out employment, or join a political party or align with a social movement or a religion even, or launch our careers based upon a voluntary response to some core idea, or principle, or belief, or premise, or aspiration or ambition even, that aligns us with the leadership of that group or organisation, or with the collective identity of that assembly – such as picturing oneself at our own graduation, or as a successful salesperson, or as concert pianist, or as an Olympian, or as a teacher or a poet, and we do this so as to generate the motivation for the sacrifices necessary to achieve those noble goals. Agreed, many of us may have less lofty goals such as simply 'making money' and some do actively choose to invest in dishonourable activities or degrading work (for various reasons to do with lack of self-esteem or because of economic necessity) but that's beside the point here, because given a choice, most of us would choose employment, and education, and careers, and pastimes that are both fulfilling and rewarding – as opposed to merely making us money – because money alone rarely fulfils that need for personal fulfilment. Enough money can distract us for a while from acknowledging that essential need – but it cannot replace it – not if you have a conscience or any real humane aspirations.

And this is where a very important principle comes in; the understanding that as we each advance through life, we are (mostly unconsciously) constantly seeking out 'that which is better' (as best we understand that notion) but, being aware of our own personal limitations in a big and often-scary world, we invariably turn to worldly institutions for guidance, and inspiration, and education, and support, and direction, and for meaning-and-purpose, not realising in the process that we are now at risk of surrendering our own individual, intrinsic values or ideals for those of the chosen institution or group. This same principle applies whether we align with a 'good', 'bad' or indifferent group, and whether we do so for good, bad or indifferent reasons or intentions.

It would be naïve in the extreme to suggest that we can navigate the modern world in complete independence without some dependency on the State or on large social institutions, because after all we are social animals and must be part of the interdependent collective. But on the other hand, it is equally, if not even more important to grasp the nature of the beast here; that almost without exception, all hierarchical institutions and organisations – and most especially governments and private corporations – exist to serve their OWN interests, NOT those of any moral, humane idealist or truth-seeker.

It is a very happy marriage indeed when a few sincere individuals gather with noble intentions and shared ethics to achieve honest goals, but even small-and-discrete groups like this will struggle to maintain a pure ethos while surrounded by 'the big boys' in the corporate playground who share no such naïve notions as they eye up the competition. Their paperwork and posters may state otherwise, and their customer charters may be models of propriety, but again, this is because the Great Lie must be obscured and 'hidden in plain sight' while most of us, distracted and convinced in equal measures by the complexity of it all, cannot actually see the wood for the trees, and don't actually understand what it is we are actually looking at.

To put it plainly, *individuals* have the capacity to develop an active conscience — one that is rooted in moral, humane values. But organisations and institutions do not. This is purely and simply (as we shall soon see) because the functionality of all large organisations depends — to a greater or lesser extent — upon the individual handing over their authority, even in limited ways, to the organisation. Where an individual has a highly developed sense of morality and personal integrity and then something 'goes wrong' within the organisation, then that's the person who speaks out. Usually we call these people whistleblowers, and usually, we then discover that the thing that actually 'went wrong' in the organisation was simply that someone with a conscience had somehow come up against the lie, and we know what usually happens next. The organisation then moves to protect itself; the whistleblower gets punished; and truth, morality and conscience be damned! In short, that inasmuch as any group or institution demands that you sacrifice your morality or surrender your conscience in order to be part of the club, then we should each be fully aware of what's happening to our own individual, sovereign authority, because once we get in the habit of giving it away...?

So, if we understand that our alignment with any particular group, or idea, or institution should only be something *temporary* to propel us to 'better and greater things' then that's fine, as long as we are directly connected (within three levels of separation) to the core goals or leadership of that group – otherwise, how can we really be sure that what we are a part of is really what it is supposed to be?

Big Pharma for example may promote itself as being driven by humane values so as to alleviate suffering and illness and to promote good health, but what is the actual truth of the matter, and what role am I playing here if I chose to work for them, or push their products? Am I really part of the health industry – or am I just shoring up another criminally-diseased corporation? Am I really a public servant who works for the common good, or, am I a mercenary tool of the political establishment, paid from the public purse to exploit and deceive the community for my own benefit? Am I really interested in the truth, in news and in ethical journalism, or am I no more than a whore of the Cabal, getting paid to bend over and continue doing the dirty as what is in effect, a furtive public-relations spokesperson of the establishment? Am I a genuine 'officer of the Court' who holds justice as a preeminent value, or am I willing to use the law and break the law in service to those who are morally and ethically corrupt?

As a member of society, have I developed my own personal skills of intuition and discernment sufficient to know what is right and what is wrong, and to see when I am being lied to? Have I chosen courage and conviction over complicity and compliance? As a teacher or parent even, am I really involved in genuine education or am I just another mouthpiece for the Cabal, making sure that trusting generations of children will be well-conditioned to the lie – and to the roles they must play if they are ever to get along and be successful? But what price such 'success' if it comes at the cost of our very souls?

We MUST, absolutely, take personal responsibility for the choices we make, in the understanding that no social institution, or group, or organisation can replace our own moral responsibilities in life. Indeed, this is how otherwise decent people can find themselves inadvertently fulfilling insidious functions whilst believing – quite sincerely – that they are simply, and innocently 'just doing their jobs'. The concept of the individual Triadic Archetype comprising, (a) one's spirit, (b) informing ones conscience, to (c) direct one's actions is replaced by the organisation's; (a) vision or mission statement, (b) the business plan, and (c) the outcome or objective which is rarely centred on humane values. If you are within three levels of separation from the leadership however, you should not only know what is *really* going on, but you will have no moral excuse to be 'innocently employed' in potential wrongdoing or wickedness. And this is why – after three levels of separation from the discussion in the boardroom, or at Cabinet, or in the war-room behind closed doors – large institutions such as multinational corporations, armies or government agencies rely on other means of control of their employees who might otherwise balk at the prospect of being engaged in massive dishonesties. They achieve this primarily through; (i) bribery and inducements (wages and promotions); (ii) via restrictive deceptions and obfuscations (*"That's above your pay grade;* [or] *take it to your superior"*); and (iii) via subtle threats and coercions (*"You do want to keep working here, don't you?"*); and all of this to be achieved through, (iv) the maintenance of strict, administrative top-down hierarchies that require our obedience and compliance.

A far as the Covid-19 phenomenon is concerned, at what point do we start questioning an official narrative that contains all of these coercions, deceptions and moral compromises so as to get us to comply? Could it be possible that lies *must* be used to garner the cooperation of otherwise honest and discerning souls – and if so, then what does this say of our discernment and our honesty if, now in knowledge of the lie, we continue blindly, to comply?

CHAPTER SEVEN

HYPOCRITICAL HIERARCHIES

& THE COST OF DOING BUSINESS

This brings us to all those other institutions and organisations that we are now so familiar with – the ones that started out as services to the people such as banks, insurance companies, health services, or medicine manufacturers, but which now exist solely to make a profit.

The medical industry is the classic example. Which is better for the bottom line; a healthy community that doesn't need doctors or medicines – or a sick community that does? Are we really in the business of 'public health' – or of that of public sickness? Which is the better doctor – the one who treats your ailments, or the one who endeavours to keep you healthy – and which one earns the most? Thus we see how and why large medical corporations or private health systems have a vested interest in having enough sick people around to keep them in business. We can also see an incentive other than the Hippocratic Oath for the burgeoning growth of the medical industry. Child protection services too started out in response to a genuine social crisis, but, once the institutional hierarchies were in place there was a need to then 'find' ever more numbers of qualifying children and broken families in order to justify the continued funding of what has in effect become the 'child protection industry' or perhaps more accurately, 'the family-*destruction* industry'?[5]

What if the very purpose for the continued existence of these industries lies in direct opposition to truly humane values? What if it could be shown and demonstrated that ANY organisation that moves from its original (humane) value-centred beginnings to a profit-or-growth driven model, can only do so by fatally compromising its

---

[5] See "THE SECRET COURTS: Child Protection or Child Abuse?" by Joe Burns.

original philosophy? Would that have us stop and think about the intrinsic value of institutions and organisations that must, by their very size – or even indeed by their very existence – operate on a profit-generating model, thereby becoming a threat to humane values, rather than a genuine benefit to humanity? [xii] What if it's literally not possible to corporatize or capitalise morality *per se*?

What if 'big business' – as a social concept – is out of alignment with fundamental humane values? What if any given business other than small 'mom-and-pop' operations that provide an essential product or service to a local community; what if the very construct of 'big business' is just well-organised and officially-sanctioned exploitation? What if governments have become *de facto* businesses – albeit very badly-run businesses that can make repeated errors without worrying about the income stream because they have in effect, a 'captive' (or trapped) market in the citizens and taxpayers? But still, each department has to justify their own particular claims to their slice of the budget – right?

The concept is relatively simple although it takes quite a leap of perspective to grasp it, and it is plainly this; that as soon as the operational policies of any organisation (even a church or a charity) prioritises the generation or seizure of funds or the making of 'a profit' – even if such is necessary to keep the project going – then whatever the noble aspirations that may have originally grounded the initiative, they now take second place (at best) to the need to make or acquire that necessary money. The project or organisation becomes a business in other words and the priorities inevitably change. This is how humane values can get displaced in the name of 'progress' or of 'fiscal sustainability' or for 'economic growth' or even in the name of 'public service' and illustrates how the first inescapable principle of business is 'expansion' (or die)! And if that expansion must come at the expense of our founding principles...?

In the case of governments or of large political parties, the dynamic is only slightly different, with the objective being administrative 'control' of a region or of a population, "for the purposes of improving society" – or at least that's the noble objective that they keep selling us in order to get their hands on the moolah, yet again. And this is how deception in its myriad forms creeps into the operational (unspoken) philosophy of large institutions whose real goal is profit, and power, and control. Those various deceptions range from false pretences to political propaganda, from individual moral hypocrisy to hidden corporate agendas, and from glitzy advertising to concealing the real truth in convoluted small-print. Everyone in the organisation is expected to participate in the lie – at whatever level they are at – or, they go find themselves another job – right? Whatever the case, the overarching consistent element in modern political institutions – especially those who are married to the private sector – is deception. But this is a covert deception that must not be seen or openly recognised by the unwitting public, who ultimately are funding it all. Nor can the public be allowed to understand where the real power lies and who is actually pulling the strings – and why. So, the lie has to be systemised, and formalised, and set in motion, and people have to be specially trained and conditioned so as to be able to present the lie convincingly – or to conceal it (from the public at least) – in some outwardly-acceptable form which, more often than not, is disguised and camouflaged under tiers-upon-tiers of bureaucratic obfuscation; upon deflective redirections to other State agencies or operators; by evasive apologetics and disingenuous legalisms; by blatant gaslighting and stonewalling and by reams of bewildering, mind-numbing red tape. This is why so many at the senior level of so-called 'public service' are actively recruited NOT for their abilities to do the job according to public service requirements, but instead because they are skilled and adept at all of the required deceptions and deflections.

Those of us who have had intimate dealings with agencies of the State are fully aware of all of the contradictions between what's written on the tin, and what's actually inside it. We also now know why civil and public servants are so very defensive and cagey when asked to fulfil their stated role as 'public servants'. Because that of course, is one of the very first lies. They *know* that they are not really public servants at all but are there to serve the Cabal – but YOU are not supposed to know that – otherwise someone might then ask why it is they are being paid from the public purse – right? Worse still, the public might actually start insisting on value for money.

In order to achieve any of this systemised general deception, governments and State-aligned corporations must have structures and hierarchies, and large ones at that, that invariably and inevitably suffer the forms of corruption we have just discussed, no matter what the original noble intentions of the founders may have been. The one interesting exemption, quite ironically, is the military, where a certain fundamental honesty underscores the violent nature of that institution and its willingness and ability to 'project lethal force' as-and-when it is needed. This requires that soldiers obey orders unquestioningly, so there is no need for the military to lie about or cover up their methods or objectives, because the objective is what it says on the tin; to project lethal force. That's also why soldiers are trained to obey orders. The great lie here is not embedded in the ordinary ranks of the military who understand that their role is to simply obey orders, because "You're not paid to think lad!" The great lie here is in the concept of 'national self defence' as a means to generate massive corporate profits for the industrial-military complex that is again, wedded to the political establishment and all of the free money that comes with that alliance. Remembering that there would be no need for massive, expensive militaries if there was no immediate potential for war, we see again how and why 'hidden hands' throughout history have invested so heavily in keeping us fearful and suspicious of each other; i.e. 'the enemy' – that terrible threat – whomever that 'other' may be.

So, If we equate the concept of 'corruption' with 'general moral wrongdoing' and with 'the pursuit of inhumane things or ideas at others expense'; and if we understand that our place in any given hierarchy (other than at the very top) invariably requires the surrender – or part-surrender – of our personal, moral sovereignty, then we begin to see a disturbing connection between the notions of 'good' and 'evil' inasmuch as the very growth and expansion of large organisations requires a corresponding retreat of humane values.

Again, the Triadic Archetype model comes into play here whereby the worst aspects of the corrupted sociopathic masculine almost invariably feature in the growth and expansion of big business. Because, if the only way to consistently 'make money' for example (or misappropriate money from the taxpayer) is through the active exploitation or deception of others (in one form or another) then we begin to see how all large corporations, organisations or institutions must inevitably become corrupted by that drive for profit, or growth, or sustainability or whatever we might call their particular justification for their own existence. We can also see how and why certain types of people with particular skill-sets are needed to populate large or growing organisations. People who are adept at deception and even self-deception in the quest to justify their jobs and generate profits or appropriate funds for the group, by any means and above all else.

You see, if money can only be made or raised by exploiting others through taxes, by deceptive marketing, by misinformation or propaganda, by threats or fearmongering, or by begging and pleading (playing on others' emotions) or even by appealing to people's social responsibilities to do one's part in time of war for example, or through any other form of trickery or theft (officially-sanctioned or otherwise) then clearly, that type of 'fundraising' is wide open to abuses and corruption. Advertising works as they say — which is why marketing and PR is such big business. But then again, so does government propaganda. Combine the two and you get a very effective (albeit deceitful) platform to generate ever more funds and induce the compliance of the public. And this is what's happening right now. Big Pharma, the media and the political world tag-teaming and scheming together to deceive, mislead and exploit the rest of us.

In this manner, the allure of the poisoned fruit is ever-so-subtly, but enticingly laid out before us in all of its insidious, institutionalised splendour; the pay rise; the promotion; the new car; personal power, wealth and authority; the attentions of the opposite sex; insider connections, titles, awards, gilt-edged pensions and public accolades and 'attaboys' from above and whatever else is needed for us to dumb down our consciences and sell our souls. And so the creeping conditioning infects and infests our consciences like a cheap but powerful narcotic. Even if they were not already knavish by nature at the outset, those who continue in these exploitative roles or who knowingly facilitate them will soon become conditioned and attuned to the sociopathic mindset – at least as far as "just doing their job" is concerned. In fact, in most corporate settings ruthless profiteering of this sort is actively encouraged and glorified, and those in government agencies who effectively obscure the great lie are similarly rewarded.

A simple but poignant real-life anecdote comes to mind, where an honest young woman takes a part-time job in a restaurant. Starting at the bottom, she gets all of the dirty donkey work, scrubbing floors and cleaning out grease-packed drains on the minimum wage. One day, the Chef approaches her complimenting her on her admirable work ethic and presenting her with an opportunity to work in the food preparation department instead – a promotion indeed. Flattered and surprised in equal measure, the young lady graciously accepts and proudly informs her family of her advancement and of the accolades of the boss. But on the very first day in her new role, the girl is instructed to change the date-labels on the stored foods so as to extend their shelf-life and ensure that the restaurant won't fall afoul of any surprise food inspectors. Technically of course, this would be an illicit and dishonest action inasmuch as food that is officially 'out-of-date' will now be served up to unwitting customers. Even allowing for the Chef's superior culinary knowledge and his own genuine belief that the food is still 'safe', there is no doubt that a deception is afoot – and a possibly dangerous one at that – and that it is being conducted at the public's expense for the primary purpose of providing more income to the business.

What is the honest young woman to do in this situation? Does she just 'follow orders' and facilitate the deception, knowing that 'obedience to the boss' is the best way to ensure promotions and pay rises? Or, does she respectfully decline the opportunity, and in so doing demonstrate to all involved that no job, or promotion, or pay rise can compensate for an untroubled conscience. Well, the clue is in the description. The 'honest' girl did the right thing and returned to skivvying – which she continued to do honestly, to the best of her ability – secure in the belief that 'making right choices' especially when they are fraught with difficulty, will bring its own reward.

Several months later the girl left the job to travel abroad. Upon her departure the Chef presented her with a glowing reference and a freshly-pressed chef's uniform, complete with the promise that should she ever wish to come back, it wouldn't be as a skivvy or in food prep or as a well-paid waitress, but in the role of a trainee Chef. I have no doubt that had she not shown such a solid work ethic and put her principles first, that she would have left that employ unnoticed and unremembered, except as just another of the train of dutiful compliants who can be relied upon to just do as they're told.

In like manner, as a young man I was offered the post of Chief Instructor at a resort in the Alps, only to find upon my arrival that the British tour company had hired half-a-dozen unqualified reps as ski instructors. The position of 'Chief Instructor' was going to sit very well on my resume, and the pay wasn't bad either, but even though I was confident that given time, I could train those reps up to a minimal standard, I was still very aware that the travel company was pulling a fast one on its customers by presenting them with untrained, uniformed reps in the guise of professional ski instructors. The holidaymakers of course were unlikely to know the difference unless they were experienced skiers. But I knew, and I was supposed to be the Chief Instructor – the person who ensured proper standards for the clientele. I was now faced with a dilemma; that of loyalty to the standards of my profession and to my own conscience, or, to the company and the position, and the money that came with it. I shared my concerns with company management advising that the only way I could accept the role of Chief Instructor in good conscience was; (a) if the holidaymakers were told the truth upon arrival, and (b) if the reps attended regular ski clinics with me to learn the basics and upgrade their skills. I knew I was putting my position at risk, because now that the ski season had officially begun, it would be difficult to secure another quality position elsewhere. But I stuck to my guns. Disappointingly, but not unexpectedly, the company let me go and gave the position to a very irritating Eton schoolboy, who, on the back of several unfavourable client reports, soon found himself out of a job too. In fact, it was that particular episode along with similar such incidents in other ski resorts around that time that eventually led to demands by the French authorities and by ski-teaching associations that *only* properly-qualified ski instructors could legally teach. For those that care, yes, I did move on to bigger and better things, but that's another story for another day.

*   *   *

This is how we slowly numb our intellects and our consciences and effectively 'sell our souls' in return for a wage, a mortgage and a peaceful retirement, and then justify our own moral decline; the loss of our courage, of our integrity, of our honour and dignity, and ultimately the abandonment of our very essence as true human beings who are intrinsically gifted with such wonderful intelligence and self-awareness, and with inalienable personal freedoms. We were clearly born superior to the beasts of the field who are absent the type of self-awareness that provides each of us the opportunity to be so much more. Yet, given the choice between a natural, moral self-determination which challenges and rewards in equal measure, or between mindless servitude in the employ of another, most of us, paradoxically opt for the latter, unaware it seems, that we are so very unaware.

In the aforementioned, *"Discourse of Voluntary Servitude"* de La Boétie remarks on this "greatest evil – this unparalleled vice" which he struggles to name. But what he was talking about in contemporary terms was the voluntary surrender of our essential freedoms; the giving over of our souls and our consciences if you like, to other forms of authority that always and inevitably descend into tyranny. Perhaps more easily identifiable in the social inequalities and dictatorial abuses of Dark-and-Middle Ages monarchies, where the longsuffering peasants just hoped-and-prayed that it wasn't their turn just yet to be raped, pillaged and plundered, so are the same dynamics still with us in slightly more subtle forms, with the establishment as always, plotting and scheming to keep the peasantry in their place while they set about enriching themselves and their cronies, through truly diabolical deceptions and through hierarchies of hypocrisy.

In fact, if we really thought about it for a moment we would soon come to realise that even though we have become thoroughly accustomed to having all sorts of social organisations and political institutions dominating human life and history, that there is a very strong argument that none of them are actually necessary to human life and fulfilment. They *appear* to be necessary to us because we are thoroughly conditioned to seeing them everywhere – including throughout history – but all of these structures and institutions exist simply because enough of us have, in one way or another, delegated our sovereign authority over to them. What would happen (one might imagine) if each of us took full ownership of our own intrinsic humanity, and moved from a position of habitual, willing and ignorant dependence – continuously waiting for 'the authorities' to tell us what to do – to a state of enlightened inter-dependence upon ourselves and likeminded others – where humane values and a heightened sense of social conscience and community guided all of our actions, our activities and our pursuits? Who would need these institutions then, other than those who would use them to exploit the rest of us?

But still you might say, it seems relatively pointless and even futile to try to elicit moral or social changes when 'the system' (the establishment) is so very powerful, and we are not. Best to just put up and shut up, and be thankful for whatever freedoms they allow us, even if those freedoms are just crumbs from the tyrant's table. "Don't rock the boat" in other words, even if you know the crew are all pirates and the ship is heading in the wrong direction – right? Because the alternative is to drown in a sea of unknown terrors – terrors which 'they' tell us, they are actually protecting us from with their deceptions, and with their 'necessary' rules and organisations.

But what if we were to tell you that you're not a mere passenger on some ill-fated craft with no say in where it's going, but that you are in fact the true master of your own destiny; if you would but believe it? What if we were to tell you that in understanding this crucial reality you touch on the very meaning of life itself? What if we were to suggest that most people travel through life without the remotest idea of what it is they are a part of? In fact, most don't even care. But 'not caring' is a denial of who we really are – or at least of who we were born to be. And very interestingly, this is not so much a religious concept as it is a scientific one – one that aligns with the 'collective consciousness' aspect of The Triadic Archetype where we all aspire to a common ambition to be the very best versions of ourselves, both individually and collectively. What if this is what 'true faith' is really all about? Not so much a belief in some external Deity who will, at some prophetic point, come riding to the rescue on clouds of glory, but a single-minded belief in our own intrinsic power to actually make a difference here-and-now, if we can but muster up the courage to try, provided we are being guided by truth and by higher moral aspirations. A personal 'faith' if you like, not necessarily 'religious' *per se* but certainly Universal, that we can manifest 'the Divine' in our concepts, in our plans and in our actions – both individually and collectively. This too requires commitment, persistence and tenacity in seeking out the truth. But frankly, without it we are utterly lost.

So, maybe it's time to challenge the traditional tyrants and predators; the liars and deceivers; the manipulators and the parasites; the corporate and political pirates who are involved in these pernicious partnerships who would steer the ship of human society onto the proverbial rocks where they can better plunder and enslave us all under the guise of a necessary rescue from the Covid-19 storm; albeit a contrived and wicked storm, of their very own diabolical creation?

Yet still. We, the fools, sit in the galleys bending our backs to the oars, doing the bidding of the diabolicals, still confident that the ship of human society will somehow survive this Covid storm – or at least, that there will be a place in the lifeboats for us if things get too bad.

But that's not the plan Folks. It never was. There is no rescue for us here – and there are no lifeboats. Our awful and inescapable destiny is to go down with the ship unless we muster up the courage now, each as free and independent sovereign souls, to take the moral plunge! Better to swim for ourselves and inspire others to do the same, than to passively assist in driving humanity to imminent doom.

Because surely, for any person of conscience, the greatest terror of all is the fear and the possibility that we too may have been an unwitting instrument of evil?

The Empathic - Sociopathic Scale
Fig 2: The Modern World

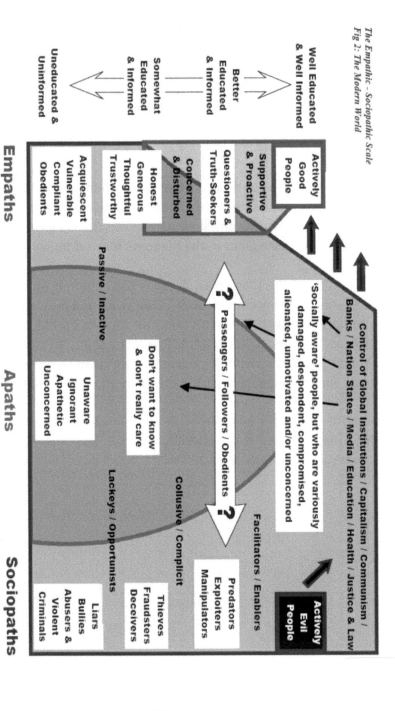

© Stephen T Manning, 2021

94

CHAPTER EIGHT

PROFIT AT ANY PRICE

THE WAGES OF SIN

In these closing chapters of Part One (before returning to the specifics of the Covid-19 phenomenon in Part Two) we examine the role and place that money – and our desperate need for it – has had in modern society and how, arguably, it has made knaves, slaves and fools out of so many of us. With all due respect to those of us who have ventured throughout life to simply 'make an honest living' and who, quite understandably therefore balk somewhat indignantly at being compared to "crooks or fools" the sad fact of the matter is that unless one owns one's own completely self-sufficient island and have never had to partake of 'the economy' then we have – either directly through the act of buying and selling, or by investing in questionable stocks and shares for example, or, indirectly by paying our taxes, contributed to one of the greatest mass larcenies ever.

I know that it seems unreasonable and even borderline ridiculous to speak of money this way when everyone knows that, "money makes the world go round" – right? Because none of us can escape its influences and its demands and, other than by living on our own island or retreating to a Buddhist cave in the Himalayas, none of us can claim that money isn't important to us and isn't right at the centre of our concerns about maintaining a certain quality of life. But that's precisely the point here. Money, in many ways, has replaced the original humane 'currency' of the natural, spirit-centred empath who gives of their time, their energy and resources purely because someone is in need, and NOT because they get paid for it. The idea of communities and societies living in a state of natural, empathetic harmony without the need for a money currency is comprehensively covered in the aptly named book, "Humantruth – A Philosophy for a World in Crisis" by the equally aptly-named author, the late John Bapty Oates [John-the-Baptist?]. When a book is described by a noted academic as, "Possibly the most important book ever written" then hopefully that is enough to prompt further investigation. Please, I urge you, read John's book. You will never look at life the same again.

Very interestingly, although John Bapty Oates considered himself an atheist and a pure scientific thinker, he nevertheless devotes much of his book to the intriguing notions of 'humantruth' and 'supra-consciousness' both of which he maintains are crucial to human development but which are sadly absent from modern societies – at least at the collective level. The result is a world of chaos and contradictions, of lost souls if-you-like who don't even understand their intrinsic capacity for what John calls 'intellation' which is a step above (mere) human intelligence or everyday consciousness. If we were to realise these capabilities suggests Mr Oates, then our world would be a vastly different place, devoid of all of the destructive aspects of human nature and culture as we know it today as well as being naturally absent any plague of sociopaths. Such a re-imagined world he argues, would also be free from any of the constructs and institutions that perpetrate our "wrong reality" and which prolong the sociopathic dysfunction. Chief amongst these ills Oates lists the inequities and inequalities of profit-driven "money economies".

What was very interesting to me though, as someone who has studied both theology and psychology, was in noting that these three novel concepts (humantruth, supraconsciousness and intellation) equate almost exactly with what we have described in this book as 'spirit or soul' or as our connecting point with Pure Truth or The Divine or whatever name we choose to assign to that Great Unknown. In other words, even a pure scientific thinker of Oates' calibre understands and acknowledges a 'special something'; something which is crucial to *true* human existence that is generally absent from the constructs of human history, and that if we are to have any hope of redemption or of salvation from 'the Machine' that we must recover or rediscover that lost supraconsciousness – i.e. the still-absent 'spirit' or 'soul'?

Not wanting to get too philosophical, because after all this is supposed to be a book about the Covid-19 phenomenon; but in the same vein being aware that we need to be able to see 'the bigger picture' if we are to have any chance of dealing with this burgeoning Covid-19 problem; we really need to *get* the fact that some of these notions and constructs are absolutely a part of that 'bigger picture'. Besides, I am aware of only one comprehensive explanation for how and why (it seems) that a vast chunk of professionals operating from within the major corporations, the political institutions and the MSM are peddling a narrative that simply doesn't hold water, and that explanation will require some basic understanding of the topics we have covered so far. Because it is no exaggeration to say that this Covid-19 phenomenon could very well be the changing point in history where we either wake up and reform ourselves into properly-functioning, humane societies, or, we all sleepwalk blindly, into Hell.

The Problem With Money. Because of the pervasiveness of money and our natural urge to want more of it (I mean, who has ever said they don't want more money?) ..so, whenever there are opportunities to get more money, we usually take it. Money is an extremely valuable commodity after all, and most of us in the developed world are fortunate enough to have enough of it to have reasonably comfortable lives. But money is not only a convenient means of buying and selling, it is also a commodity in-and-of-itself, and there are people such as bankers, stockbrokers and hedge fund managers for example, as well as large financial institutions, corporations, and governments involved in Foreign Exchange Markets who, because each of them are in the 'business' of accumulating as much wealth as they can for themselves, for their clients or their respective countries, create the circumstances whereby unfortunate others (such as those in Third World countries) are deprived of enough money to even live.

Speaking purely from an empathetic position, it seems terribly unfair that some individuals and some nations have more money than they can burn, while other people are dying from a lack of it. Not because of any unwillingness on their part to work and earn a living, but simply because of the circumstances of their birth and the poverty they have been born into. One would think that a species as intelligent as us would have the wherewithal to come up with a more equitable system of exchange and commerce than a money economy which fosters overt greed and rewards cruel exploitations? But apparently not. In the same vein however, what's the alternative? Do we return to a highly-impractical bartering system or should we be thinking of creating some new merit-based system of exchange that gives equal opportunity to all? Well, I supposed it all depends upon your perspective and upon whether you think like an empath or a sociopath – right? The empaths will want a fairer system, whilst the sociopaths will not. They will still want their profits – at any price!

Now if you are anything like me, you will shake your head in equal measures of wonderment and bewilderment, and in general appre-hension and dismay at the internal goings-on of world economies and at the convoluted intricacies of banks and stock exchanges. You might also struggle (like me) in grasping the general concept of 'wealth' or of the intrinsic 'value' of any given thing in monetary terms. Because who sets the value of land, or of property, or of various commodities for example? Or perhaps far more importantly, who sets the value of the almighty dollar – the base currency for all modern economies?

Briefly, we think back to the confusion and puzzlement of the Native American Tribes when first approached by the White Man with offers to 'purchase' their land, with Chief Seattle sagely responding: *"How can you buy or sell the sky? The warmth of the land? The idea is strange to us. The Earth does not belong to us, we belong to the Earth. If we do not own the freshness of the air or the sparkle of the water, then how can you buy it?"* But this hasn't stopped 'the white man' in particular from insisting that the 'worth' of anything can be quantified and monetised no matter what the cost may be to humane values.

Now we're not claiming to have all of the answers here, but in considering that one major aspect of the planned 'New World Order' is to have a cashless society, then it may help us to have a general understanding of what it is exactly that 'they' need to replace, and why the apparent need for such a radical change anyway? Because interestingly, one solid theory about the arrival of the Covid-19 phenomenon is that it was never about any genuine 'heath crisis' at all, but that, "this was always about the economy folks"; a looming financial crisis and a massive power-grab that no-one at the top wants us to know about, or understand. Because quite frankly, if the general population were able to get their heads around the incredible and indeed obscene levels of greed and mass-exploitations of the public that has been going for decades 'in high places' and of all of the insider-trading transfers and seizures of genuine collective wealth by the Cabal at the expense of true humane values, then arguably, one of the first financial investments of any community-driven 'New World Order' with a properly functioning justice system, would be to bring back a highly efficient modern form of Madame Guillotine.

You see, at one time, a community's 'wealth' was measured in its local resources, and on how well it could support itself without going elsewhere for the stuff it needed to survive such as food, water, building materials, and labour for example. But it wasn't long before the leaders of those communities (usually the strongest, most ruthless-and-ambitious [sociopathic?] types) came to the fore, and, as all well-adjusted psychopaths are wont to do, they set about organising raiding parties to forage afield to rape, pillage and plunder and thus steal what they wanted from other communities. But this could be costly in terms of lost manpower – not to mention that it was bloody hard work (excuse the pun). The more empathetic leaders (and the less powerful psychopaths) would have no doubt seen the immediate mutual benefits of trade and barter (vs rape, pillage and plunder) and this is how (very simply) the idea of using inanimate objects such as precious jewels, or rare spices and perfumes that had no essential value to the community in terms of providing food and shelter for example, became tokens of exchange for things that *did* have essential value. As communities advanced in sophistication they then started making ornaments and jewellery from these precious objects and soon discovered that the two most malleable metals (silver and gold) made very nice shiny things that you could stick jewels and precious stones into as well, and, because these items were easily portable, they became symbols of individual wealth and power that could be easily displayed or used in trade. Unfortunately, this also made it easier for unconverted psychopaths such as the Vikings (apologies to our Nordic friends) or to various warring emperors and monarchs, to acquire great wealth through warfare. This notion of 'warfare equals progress' is *the* cardinal giveaway that any particular society remains trapped in the wrong dynamic.

The brilliant idea that someone then came up with in 7th Century BCE Turkey (we believe) was to strike easily-portable coins of silver and gold, and mark them with values to make trade easier. When the idea caught on, others started adding national symbols to protect any given nation's internal resources (to a certain extent) thus creating the world's first money economies whereby both essential goods and resources, as well as luxury items, were purchased with— or replaced by—gold and silver in ever-expanding cycles of trade. In this manner, the discovery of gold and silver on one's land, and/or the acquisition of it through trade or warfare were the singular means by which any nation or community could *directly* increase its intrinsic 'value' other than by the wholesale seizure of land and the appropriation of goods such as we saw with the British East India Company who, very interestingly, also struck and minted their own coinage to facilitate international trade. In fact, this was also a key driver in the white man's pursuit of Native American lands around the same time period. The eternal quest for gold or indeed for anything that could be traded for it is well written into the history books, along with all of the deprivations and wickedness that accompanied it. And, at the risk of pointing out the obvious, if we think of all of the things that humankind has done in the quest for ever-more power and wealth, well, all that's left is for us to be truly ashamed of ourselves.

The history of money, and of debt, and of accounting and national currencies is convoluted and mindboggling in equal measures, and the explanation above is just a bit of a tongue-in-cheek attempt to bring the non-financially-trained reader (such as myself) to a point of basic comprehension as to what money actually is, and how it works today.

So, having reached a point in history where the nations of the world could now trade with each other using the very convenient medium of 'money' in the form of metal coinage; more convenient 'notes of exchange' – the precursors of modern banknotes – began to be used between traders instead of carrying huge amounts of coins for major transactions, and these notes of exchange would be guaranteed by the respective sovereigns, governments, banks or corporations. These credit notes in turn could be redeemed for gold, and that gold was kept in national vaults, or banks for safekeeping. But obviously, countries could only print out otherwise-worthless paper credit notes up-to-and-including the value of their own gold reserves.

This, for want of a far more complicated explanation, is how we got what was known as 'the gold standard' whereby in the interests of creating a stable worldwide economy, countries around the world agreed only to print as much money as they had in gold reserves, and they all agreed that one ounce of gold was worth a specific amount. Between 1834 and 1933 for example, the price of gold was set at $20.67 (US dollars) per ounce, meaning that if a country such as Switzerland had 100 tons of gold in its vaults for example, then it could print Swiss Francs (at the respective exchange rate to the US dollar) up to the value of that 100 tons of gold. This 'gold standard' concept is extremely important in understanding what is going on in the world's financial markets today, because obviously, if any country that was on the global gold standard started printing more money than they could cover with their own gold reserves, then that would lower both the value of their own banknotes, as well as compromise the internationally agreed-upon gold standard price of $20.67 an ounce. In simple terms, if Switzerland were instead to print Francs to the value of 200 tons of gold and then everybody tried to redeem their banknotes for physical gold at the same time – well, there just wouldn't be enough gold to go around, would there, and half of the creditors would get nothing at all. So it was important that everyone honoured their part in the scheme, or else the global economy could collapse in a galloping panic. (I hope everyone is following the logic).

At the Bretton Woods Conference in 1944, the 44 Allied Nations renewed the gold standard at $35.00 per ounce, acknowledging the position of the US dollar as the world's leading currency. USA also held by far, the largest stocks of gold at Fort Knox, and so for a while all seemed to be well with the world. But then the USA launched the NASA Space programme and got tangled up in Vietnam, and these were both very expensive initiatives to fund. Some countries started getting nervous about how much money the Americans were printing and spending and about the resultant drop in the values of their own gold reserves (some of which were held in trust in Fort Knox) if the Americans kept on printing money, and they started asking questions. The result was that in 1971 President Nixon placed a 'temporary protection' (which soon became permanent) that un-linked the US dollar from the gold standard. Instead, he simply promised all the countries in the world that they could "absolutely rely on the USA" to pay any debts they incurred, simply because the US had the strongest and largest economy in the world. Everyone believed Richard Nixon of course, because as we all well know, politicians as a rule do not lie!

In taking this action the US dollar became a "Fiat" currency meaning 'by Government decree' which basically means that the Central Bank can print money for the government anytime it's asked to – and can do so (theoretically) without any gold bullion-related limits on how much money is being printed. But clearly, the only responsible way to be printing new money – just as with the old gold standard – is to have something of value (such as more gold or resources) to qualify the production of the new money – right? In other words, if a country like Norway for example bases its own Fiat currency on the amount of trees it exports every year, and then it produces 10% more trees the following year, then it would be understandable how they could justify printing 10% more Kroner (or Euros) without devaluing the money already in circulation. In fact, this is how inflation is supposed to be controlled. By only having funds in circulation that accurately reflect the value of one's national resources, including exports, labour and services for example.[6] But obviously, there are consequences for the over-printing of paper money, because each new Dollar or Kroner or Euro being printed *without* the requisite value to back it up actually *devalues* the money that's already in circulation – correct? And that, in short is what has been happening for decades now. The fact that gold (at today's rates) will cost you a hefty US $1,800.00 per ounce should tell its own story.

Now the last part of the financial puzzle is in understanding where the USA is getting its extra funding from? And what is it that allows the Federal Reserve (USA's Central Bank) to keep on printing more-and-more money to keep the American economy blazing along without going totally bankrupt? Not forgetting that the USA remains the world's largest economy, I must admit that I was alarmed to hear that the US Government's main source of funding comes from the sale of US Government Bonds. In essence, this means that the US Government is issuing IOU's to their own Federal Reserve (Central Banking Authority) in the form of paper bonds in return for new printed money. These IOU debt-bonds are then purchased by outside investors (including other countries such as China, Japan and Ireland for example) who trade them on the open markets on the simple basis that the USA will always pay its debts, and upon which confidence, the debt-bonds and the interest the bonds are

---

[6] [Making money from direct speculations on money-related products is a far more complicated discussion that we need not examine here.]

generating for those in possession of them, are as good as gold so-to-speak. The really big problem here though, is what happens if market confidence is lost and no-one wants to buy or trade those bonds anymore? What if a global health crisis for example, undermined the US economy sufficient to raise questions as to the future value of US Bonds? This is a very real prospect as-and-when everyone involved realises that this Fiat currency concept is no more than a massive ponzi scheme that simply cannot continue forever. Especially not when US debt is now running at an alarming $30 trillion dollars with its real-time debt-clock spinning at startling rates.[xiii]

If we are to believe the financial experts; that financial collapse is imminent, and along with that pending collapse (just like in 2008 – only far, far worse) will go everyone's hard-earned savings and pensions. Everyone that is except for the super-rich; for those who are betting on the global financial collapse; and for those in the employ of the State of course who have guaranteed wages and pensions, which is why we are predicting that the rage of the people at such a prolonged betrayal by 'the establishment' would be such as to anticipate a very bloody response. Or at least, perhaps it should!?

Unfortunately, by the time most of us wake up to what's really going on, we will be so desperately indebted and/or so utterly dependent upon the beneficence of the State, that we literally won't be able to *afford* to do anything about it.

Underscoring how those who control the money effectively control the world, the founder of the Rothschild Banking dynasty, Mayer Amschel Rothschild said in the late 1700's: *"Give me control of a nation's money and I care not who makes its laws"*.

You see Folks; it's always been about the moolah!

CHAPTER NINE

LIVING BEYOND OUR MEANS

THE TRUE COST OF DEBT

Our dependence on debt – individually, collectively, nationally and globally – also plays a big part in the equation, and that too needs to be understood in context of what is happening with Covid-19. The plain fact of the matter is that nations and governments all around the world have been spending far more money than they are generating, and over a period of many decades now, the levels of debt have become almost impossible to manage in any realistic way that does not project generations of future workers into lifelong poverty, as the interest rates on trillions of borrowed dollars, pounds, euros and drachma, reaches truly unsustainable levels. The four obvious questions to be asked are: (i) Why are we borrowing so much? (ii) Who are we borrowing it from? (iii) How are we going to pay it back; and (iv) what happens if we don't (or can't) pay it back?

The best way to understand the situation is to picture an analogy where Mr Bill Melater decides that he'd like to start his own one-man business selling ice-cream, and he borrows $10,000 from his friend Mr Robin Banks, to purchase a second-hand van, the equipment and the stock; understanding (for simplicity's sake) that he will have to pay back Mr Banks $1,000 per month for a year, making a total of $12,000 in repayments. The interest on the loan is therefore $2,000, or 20% of the original loan, and the business itself is 'worth' $10,000. So, having spent the whole of the $10,000 loan on the start up, Bill must of course now sell at least $1,000 worth of ice cream during the first month (plus cover his operational costs) if he is to pay Mr Banks on schedule. But the first problem we run into is that Bill only started the business because he likes driving around and going to interesting places, and this is more important to him than selling enough ice-cream to pay the bills. So, when he can't pay Mr Banks at the end of the month, he goes to the great Mr White (who is a bit of a shark) and borrows another $1,000 from Mr White to pay off Mr Banks. But now Mr White is expecting his $1,000 back the following month, plus his 20% interest as well. But Bill, seeing how easy it is to simply keep borrowing from other sources against the perceived *value* of the business, instead of concentrating on selling enough ice cream to stabilise the business; well, he just keeps driving around enjoying himself until he runs out of people who are willing to lend him more money each month, and all of a sudden he has multiple massive debts that he cannot possibly pay off even if he sells off all of his assets.

But interestingly, the debtors don't necessarily want Bill to sell off the business, not unless that's the only way they can make a profit because as long as the business exists, the interest is piling up, and that means that someone associated with that business is somehow going to have to pay, sometime! Because after all, the banks and big bondholders are not a charity, are they? They could of course just bankrupt Bill and sell off his assets (as they often do) but sometimes there's more profit to be made keeping someone in eternal debt.

"What!?" say the banks. *"Did you think we're in this for the common good? Or that we should warn you when your debt is getting out of hand? Or just forgive the debt as a gesture of human kindness? Do tell us Sir – how would that be profitable?"* So Bill does the only thing left open to him; he signs a contract committing his children and grandchildren to indentured servitude for life. The fact that the creditors have more money than they could ever spend doesn't even come into the equation – nor does any concept of compassion or altruism, because you can't be a successful moneylender unless you are already quite some way down the sociopathic scale – right?

If the reader hasn't already worked it out, 'Bill Melater' is America, or any one of the highly-indebted first-world nations who have been recklessly spending-and-borrowing for decades beyond their means. Mr Robin Banks is the Federal Reserve, and Mr White represents all of the other investors and bond-holders who (we are told) "are too big to fail" and whose interests (literally) are THE priority over everything else, because at the end of the day, it's all about the moolah – right? That's why we *earn* interest on monies that we deposit in the banks, because the banks re-invest that money in ice-cream business start-ups for instance and then reap their 20% in returns, which they then kindly share with us grateful depositors at lower rates of course. When we want to *borrow* money however, it's going to cost us in interest because the depositors (and the banks of course) MUST get paid their expected percentages. Otherwise what's the point of having profit generating banks that thrive on the exploitation of their customers and clients in the first place – right?

Please bear with us here, because the other very important issue to consider is what would happen if Mr Banks stopped claiming interest on Bill's debt? You know, that rather unbelievable scenario whereby Bill could borrow money from Mr Banks and from Mr White and NOT be charged any interest on his loans? In other words, that Bill *only* has to pay back the exact same amount that he borrowed? Now wouldn't that be an amazing, altruistic development by institutions that we had previously thought to be utterly ruthless and corrupt? Naturally that would require Bill to improve his work ethic somewhat and actually earn enough money to pay back his debts, but the idea of banks lending us money out of the goodness of their hearts and NOT making a profit on it? Ah now, let's not get carried away with ourselves!

Well Folks, let us enlighten you to an even more astounding development which, believe it or not, you have probably already been notified of via correspondence from your bank and probably didn't give a second thought to: That you (the investor / account holder / depositor) will have 'negative interest rates' applied to your deposits from a particular date. "Negative interest rates" you say? That sounds interesting? But what does it actually mean? Well, it means that the banks have decided that they are going to *charge* you money on your deposits with them – instead of paying you interest as normal. So, if you had $1,000 in the bank last month, then this month you will only have $990 because the bank has applied a 1% 'negative interest rate'. This means that from that particular date any money you had in the bank is not only NOT accumulating any interest, but that you are in fact being charged and penalised for leaving your money in there, while the bank, in basic terms, steals the value of your money bit-by-bit, thereby increasing the value of its own internal holdings.[xiv]

So, are they *lending* money at negative interest rates as well you might ask? Yes, believe it or not, some of the banks are, such as in Germany and Denmark. This opens yet another very important, logical and commonsense question about the whole nature of current economics; because if Bill can borrow money from so many sources at any given time (including from banks who are now offering negative interest rates) then why doesn't he simply borrow whatever massive amount he needs to, so as to clear ALL of his debts in one go, safe in the knowledge that the negative interests that are being applied to his loan will eventually, and naturally over time, actually eliminate whatever debts he has!? For example, if Bill can find a bank that is offering negative interest rates on loans and he borrows the $12,000 dollars he needs to clear his original loan, knowing that he only has to pay back $11,900 at the end of the month for instance; and then the following week he borrows that $11.900 from another bank to clear the previous debt, knowing he only has to pay back $11,800... and so on. Well, it's an obvious way to profitably fund any enterprise – even stupendous national debts of multi-billions – right? So, why isn't everyone jumping on this negative interest rates bandwagon like piranhas at a waterpark? Because surely, if there is some mechanism in place that could eliminate all debt overnight, then the governments that care about us so very much would avail of it? Wouldn't they?

Something just doesn't make sense here. Banks have NEVER ever just given away money for nothing, which raises the question of what's really going on here? Because clearly, there has been no desperate rush to eliminate global governmental debt via these negative interest rates. It's almost as if something else may be going on behind the scenes, because who exactly were we borrowing all of this money from anyway all these years, if not 'the banks' (or their unseen backers) and why on earth would they now move to implement negative interest rates if not to force everyone to withdraw their deposits from the banks and stuff their cash into their mattresses? Which raises the next important questions: (i) Why would callous profiteers want us to do that? And (ii) what happens to all of the recoverable physical cash that is then withdrawn, especially if we then go cashless?

Well, one remote possibility is that at a prearranged time all of our bank accounts will be rendered fully 'digital' (card / swipe payments only etc.,) and cash-in-hand will become obsolete as official tender. There may be some grace period where we can re-deposit our mattress cash back into our bank accounts, but after that, physical cash will be worthless as official tender. This gives us the dismal choice of re-depositing our cash at a loss under negative interest rates in the bank or, risking losing it all when the transition to digital is made. Then, at the precise moment that all of the *unreturned* mattress cash is wiped out of existence, the pre-existing *value* of that now-obsolete cash will be transferred to the Cabal / the Banks / the State / the Cartel in the form of digital currency which will then magically reappear on their own balance sheets, along with whatever other invented value-numbers the Cabal decides to unilaterally assign to itself. This inflationary action of course would have the same effect as previously printing too many dollars against insufficient gold, whereby the value of everyone else's money automatically goes down. In other words, at very best, our cash-in-hand – the currency of normal trade – will be rendered obsolete, and any cash already in our accounts will be digitalised so as to 'fix' its value and transfer effective control of our accounts to the Cabal. In other words, the ONLY way to buy and sell will be through digital transactions that can be tracked, traced and indeed stopped by the State at will. Whichever way we look at it, it's a monetary win-win-win situation for the Cabal, and a financial lose-lose-lose situation for the rest of us. The proposition that everyone (except the Cabal and its stooges of course) will be placed on some minimal universal social payment that could of course be stopped at any time by the authorities, was ridiculed as 'unworkable and communistic' when it was first proposed by farsighted globalists. But as we saw with the emergency Pandemic Unemployment Payment (PUP), and now the roll-out of $1,000 monthly payments to Californians for example – in a model that will no doubt spread to other US States – the move to complete financial dependency on the State is now well underway.

Now, if I were one of the bad guys who had made a career out of ill-gotten gains through the criminal mismanagement of public resources or by profiteering on State-sponsored ponzi schemes or through other forms of financial trickery or deception; and if I was watching this potential fiscal horror show unfolding (the impending debt-driven financial collapse and unilateral transfer to digital currency) with no apparent escape-hatch by which me and my avaricious buddies might avoid being held to account for decades of unbridled greed and self-indulgence; then I would be hoping and praying for some social catastrophe to occur; some great and distracting crisis that would cause immense panic and fear that would draw the public's attention away from what we've been at all these years. Better still if we could blame what's happening back onto an ignorant public, or on the global health scare and all of its associated costs, and on people not listening to the authorities and not getting vaccinated... and then make a financial killing (excuse the pun) at the same time.

Given the likelihood that the world economy is about to come crashing down around us, the ideal distraction would be some global crisis that will not only provide a plausible excuse for the collapse of the world economy as well as a credible explanation for massive indebtedness that now requires a 'Great Reset' of the global economy (meaning essentially, that we lose personal ownership of everything) but a crisis that also provides an opportunity for the banks, nation-States and major international corporations to replace all of those pesky 'mom-and-pop' small-and-medium businesses with more efficient alternatives such as those that can be provided by the global Cabal. Or at least, to place those small-and-medium sized business in eternal, debilitating debt to such an extent that they are effectively eliminated as going concerns, other than as debt-laden contributors to the coffers of the Cabal.

It would also be helpful if that health-and-financial crisis provided us with some mechanism by which we can identify the compliant dupes and the fools, and separate them from those troublemaking cynics and questioners, and use the naïve trust of the dupes in the system to corral them unknowingly and thereby better train them for the world to come – one where "obedience to the system" will become the new fascist First Commandment. Everyone will *have* their place and *know* their place in these new hierarchies which will eventually be rolled into the New World Order where money (as we have known it) will be replaced by a global digital currency that can of course be controlled, deposited or cut-off by governments or banks at the push of a button or at the whim of the Cabal. Who knows, perhaps we will be able to inject a few initiatives into the narrative such as withdrawing State benefits from non-compliants or restricting travel, advising the public to 'get with the program' for the sake of everybody's heath, because after all, we (the establishment) know what's best. Those who behave themselves and show due deference to the system will be salaried with a living wage or, if they are especially adept at the mechanics of deception and oppression, will find themselves being offered a position in the system to be promoted and rewarded accordingly.

The cynics and questioners meanwhile, along with the whistleblowers and truth tellers will get the same treatment as any other non-compliants. First they will be ignored, censored and suppressed or, if they persist, they will be ridiculed, vilified, alienated and excluded from the 'privileges' being bestowed on the knaves and the dupes until finally they get the message that resistance is futile. 'Crucifixion' of course (either literal, psychological or metaphorical) will always remain on the table. If the crisis can also somehow tidy up our communities by getting rid of a lot of those costly old-and-infirm individuals, and thereby free up pension funds, that would be helpful too, but purely from a resource-management point of view of course.

It would be beneficial too if the crisis provided 'us' (the establishment) with some means of easily registering the global population and categorising people according to their usefulness in the New World Order, because obviously, it would be best if we only had the most productive and useful citizens availing of the beneficence of the State, because the effective removal of the economic middle class during the opening months and years of the crisis will have left only the globo-corporate elites (us) and a more-efficient, reliant and obedient peasantry (the rest of you) to face into whatever brave new world is being planned.

*   *   *

Thus, not only will 'the crisis' transfer practically all of the control over the world's resources into the mitts of the Cabal, but it will reduce the rest of us − whether compliant or not − to utter dependence on the Cabal as well. And that (literally) cannot be a 'good' thing.

We will wrap up this disturbing discussion at the end of Part Two, after reviewing the relevant data and facts. But I hope the reader is at least attuned to the possibility that in a world that is clearly being controlled by a network of sociopathic elites, by a diabolical Cabal and by the toadies and sycophants that serve them, that there are certain prospective outcomes to the Covid-19 phenomenon that absolutely must be considered if we are to have any hope whatsoever in actually doing something about it, should it become clear that all is NOT well.

THE GREAT LIE... (so far)

- That governments work for the people

- That all people are equal under the law

- That public servants serve the public

- That democracy replaced tyranny

- That political parties are not autocratic and self-serving

- That we have a genuine 'separation of powers'

- That the authorities can be trusted

- That we have the Rule of Law

- That fascism has been defeated

- That big business is necessary

- That true value can be priced

- That humane values can be corporatized

- That structural hierarchies are healthy – and necessary

- That bigger is better

- That Big Pharma is in the public health business

- That the mainstream media is objective and independent

- That 'Deep State' social-control initiatives do not exist

- That we are not all being taken for fools by the Cabal

- That we are powerless as individuals

- That 'just doing my job' is an excuse

- That staying silent and inactive is a defence

- That human society is essentially, and actively 'good'

- That justice, through law, is guaranteed

- That truth can be institutionalised

- That indoctrination and conditioning equals true education
- That cognitive dissonance doesn't affect me
- That it's not my responsibility
- That I have no sovereign, moral authority
- That my personal choices don't matter
- That we can do what we like, without consequences
- That evil doesn't exist
- That there is no international Cabal of sociopathic diabolicals
- That I am not just a resource to be used, abused, preyed-upon, discarded or removed... by globalist despicables
- That ethics trump profits
- That morality defines the professions
- That Universal Law is just a theory
- That everything is subjective
- That evil is necessary to counterbalance good
- That universal symbolism is a silly superstition
- That the Earth belongs to us
- That nature is inexhaustible
- That good will prevail, anyway
- That I have no soul
- That human evil is inevitable in society
- That "I don't *want* to know" means, "I *don't* know"
- That it will all work itself out
- That it's not up to me
- That there is no Great Lie

PART TWO

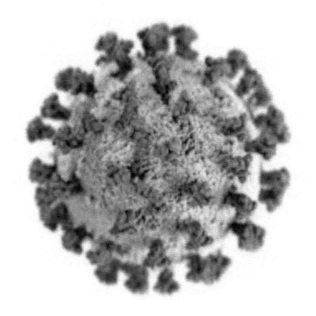

FACTS, FIGURES & FALSEHOODS

2021 UPDATE & IMPORTANT PREAMBLE TO PART TWO

In the months that it has taken to research and write this book, much has changed in the Covid-19-related landscape, so much so, that the original position that we held at the beginning (to put all of the competing narratives to the test so as to arrive at the most likely outcomes) has become somewhat redundant in light of the avalanche of published materials and incontestable proofs now available that explore and discuss many of those competing Covid-19 viewpoints in more than adequate ways; sufficient for us now to only have to clarify certain glaring anomalies in the official narrative and then context-ualise the proven facts, to be able to arrive at the shocking truth.

One hugely-important, and truly game-changing development was the publication of, "The Fauchi Covid-19 Dossier" in July 2021 by Dr David E Martin, founder of M·CAM Inc., an organisation which, amongst other things, runs a searchable index of global patents. Amongst other startling facts in that 205-page document which has been "prepared for humanity" and published for free, the dossier notes that that some 5,111 proprietary rights have been granted around SARS Coronaviruses, with 73 further patents affiliated *specifically* with SARS-Cov-2 being registered in the past 20 years.[xv]

To paraphrase these facts in more direct language; this means that SARS-Cov-2 has been 'a thing' for at least 20 years and that a number of individuals and corporations have long since known about its existence and indeed have laid claim to its creation, to its testing, and of course to its prevention and cure many years in advance of the purported 'discovery' of the novel (new) highly infectious coronavirus SARS-Cov-2 which, if we were still to believe the official narrative, emerged without warning in Wuhan, China in November 2019. Obviously, inasmuch as this definitively answers many of the questions about the true origins of the virus while eliminating other considerations, this changes the whole tenor and direction of our investigation whilst reinforcing with shocking authority the concept in Part One that we may be embroiled in something truly sinister here. Something that we now need to understand and deconstruct through meticulous forensic scrutiny, and by courageous, collective action.

To this end, we will continue in Part Two with an exploration of the various claims and purported 'facts' that were being put out by the establishment – and which continue to be broadcast as a basis for the continued lockdowns; the need for tests, masks, and mandatory vaccines; and all of the other restrictions on our fundamental rights including the utterly unconstitutional legislation that was rushed through the Irish Parliament on July 14th 2021 without any proper democratic debate, which effectively segregates the Irish public into those who have been vaccinated and those who are not.

Unsurprisingly, those who have NOT been vaccinated or who cannot prove via official documentation that they are 'Covid-recovered' are being excluded from various social venues and services, whilst the Minister for Heath conveys upon himself in that very same legislation, the unconstitutional power to draft additional discriminatory laws any time he so pleases. It's one of the oldest legal tricks in the book; to conjure up 'emergency legislation' ostensibly 'in the public interest' so as to provide an introductory vehicle for ever-more restrictive laws which, in the ultimate irony, then serve to remove the fundamental human rights which our Constitution (which is supposedly superior to those laws) was designed to uphold and protect.

With a politically-appointed judiciary waiting in the wings to shrewdly bat-aside any constitutional challenges to this blatant and illicit power-grab, and confident in the general ignorance of the public as to all things legal or legalistic, this is a classic example of how and why we should never surrender our authority to amoral elites.

So, is this genuine democracy? Or is it just tyranny in another form that is being delivered with a plastic, political, condescending smile?

117

With Dr Martin's explosive and thoroughly-researched report now at hand, it is increasingly difficult to qualify, justify or explain how and why the various authorities are continuing down this repressive and oppressive track except in context of an understanding that there is indeed a global and truly diabolical plan to achieve certain insidious objectives, which we will shortly be looking at in more detail.

In the meantime however, and with regard to all those sincere people out there; family, strangers and loved ones alike, who remain confused and bewildered in equal measure at all of the apparent lies and contradictions, the half-truths and false reports, the supposedly 'outlandish conspiracy theories' and conflicting scientific reports, or, who remain trapped in the official narrative, still convinced that we can trust 'the authorities', the following fact-and-truth based clarifications that are delivered on the basis of the information and perspectives in Part One should help to dispel any doubts that remain as to what the Covid-19 phenomenon is really all about.

We cannot remain ignorant and inactive, and clearly, 'the authorities' (or a great many of them at least) simply cannot be trusted, because they are NOT driven by conscience and concern for their fellow man. Too many have arisen to positions of power and authority in these hypocritical, publicly-funded hierarchies precisely *because* they are ruthless, devious, scheming, greedy and clever, and because it has become second-nature to them to deceive, and to exploit, and misuse and abuse anyone and everyone who is not wise to their game.

That 'game' is one where the wicked will routinely cheat, deceive, or defraud in order to win at any cost. But when that cost includes our children's futures – indeed their very lives – well, then the cost has become unacceptable, and unbearable. Those of us who now know and understand this have a duty to act because we are all in this together. Time is short, and common sense it seems, is not all that common. We can but hope and pray that we are finally doing our bit!

CHAPTER TEN

UNDERSTANDING THE NARRATIVE

AND DECIPHERING THE TRUTH

We made a promise from the start that we would only include proven, checkable facts in this book, and that is still the goal. But we must also be pragmatic. Because one well-intentioned person (such as myself) who is acting on such a strict timeline and who is desperately trying to keep the page count to a manageable level, cannot possibly research every single bit of information to the nth degree in the same way that a qualified doctor or virologist can. We need to remember that we are still in the relatively early stages of 'dealing with' this crisis, and there remains the hugely problematic issue that even if we could fully trust what the establishment was telling us (which we can't) that "the official data" on many aspects of the Covid-19 phenomenon is piecemeal, confusing, contradictory or otherwise incomplete. Not to mention the discombobulating effect of thousands of paid fact-checkers doing their utmost to eviscerate any commentary that dares to challenge the official narrative.

The net result is that hardly anyone knows where to go for 'the truth' anymore because the normal internet channels – also bought-and-paid-for – are doing the same type of filtering and censoring as the mainstream media. The ideal situation would be to have ALL of the relevant data in before firing out terrifying statistics to the public, and/or, to ensure that the public understood for example, that when the various governments started their PCR testing programs and were coming up with alarming infection rates, it was simply because they were 'testing' people (if we can really call it that) who were already symptomatic with *some* type of viral disease. So naturally, because they were only testing sick people, then no matter what form of test they applied the incidence rate between testing and 'positive' viral infections was going to appear extraordinarily high. Whilst on the other hand, if the same tests were carried out on a random section of the community instead of just very sick old people for instance, then naturally, the figures would have dropped dramatically; perhaps even sufficient to remove any need for alarm in the first place? And this is working under the assumption that the tests themselves – the PCR tests – were fit to the task that they were being used for, something we now know for a fact (if we are to believe the test's inventor), they most definitely were not! But apparently no one in authority seemed to think this was important. Or perhaps it would be more accurate to say in hindsight that those who were qualified to have an informed opinion that contradicted the official narrative and who were willing to speak out, were quickly censored and silenced – and any such truths they wanted to share were suppressed. A colleague who is a general practitioner explained that 'the powers that be' in the medical hierarchy had made it very clear that any doctor who did not go 100% with the official narrative, would soon find themselves being 'disciplined' up-to-and-including losing their licence to practice.

In any event, we now know that whatever it was they were testing for, it wasn't being done properly according to normal medical test protocols, and the resultant data that was being fed to the public was utterly unreliable as any sort of proper medical diagnosis – or as valid statistics. And whilst we must allow for the reality that many of those conducting the PCR tests were doing so in all sincerity, confident in the information being provided by the authorities, and personally convinced that they themselves were providing a genuine service, we also now know that we were all being deliberately misled and deceived by certain key figures and agencies on an almost unbelievable scale – something they seem determined to continue.

Obviously, as we go through what we know now, some eighteen months after the crisis started, we have the benefit not only of hindsight, but of a lot of credible facts and data from qualified, responsible sources. This will help us examine whether we are looking at truly horrendous incompetence and stupidity on the part of many so-called 'authorities' or, whether so many of the inconsistencies and contradictions facing us now were due to criminal recklessness or if they were the result of deliberate, preplanned lies for one reason or another?

Alternatively, were these inconsistencies the results of understandable human mistakes and of mere 'errors of judgment'-something we do NOT want to be repeating of course, as we endeavour to get at the truth in this book?

So, other than that data which I have personally researched and 'fact-checked' myself, there is a relatively simple formula by which to assess the motives and integrity of the other sources we are quoting, and it is simply this: (i) Does this person (or source) have any vested interest in saying this-or-that? If so, what specifically is that interest and does it create a conflict of interest, or raise any doubts as to that data's accuracy? (ii) What is their track record and can they be trusted? And (iii) does what they have to say align with the evidence? Not forgetting of course our evidence in Part One that by-and-large, sociopaths and psychopaths operate to a different agenda and the only real trust we can have in them, is that they WILL undoubtedly perform to type. By using this formula I hope you'll agree that we have the right approach to discern what exactly 'the Covid-19 phenomenon' comprises, and how we should respond.

This brings us to one notable example of a credible and courageous whistleblower whose credentials and motivations for speaking out against the Covid-19 phenomenon surely cannot be questioned – except of course by desperate fact-checkers and debunkers who are in the pockets of the Cabal – and yes, when you check the following quotes online, you can expect the now-usual volley of rebuttals from all of the usual suspects. (More on this in Chapter 15) Anyway, this whistleblower's name is Dr Michael Yeadon, a former Vice-President of the now-infamous biotech company Pfizer, who served as their Chief Science Officer for 16 years. Bearing in mind that some of these arguably 'prophetic' comments (in the literal sense – meaning 'of the truth') date back to the early weeks and months of the pandemic, and, other than the question mark over the precise definition of what SARS-CoV-2 comprises; the facts and data in this book absolutely and unequivocally endorse Dr Yeadon's position. We therefore open this attempt to understand the official Covid-19 narrative and agenda by listening carefully to what this courageous man of integrity has to say.

Some of Dr Yeadon's quotes include:

- *"The UK Government has been lying to the public throughout the entirety of the Covid pandemic and it's up to the masses to stop them."*

- *"The PCR test is not a valid diagnostic tool and should not be done on the scale we're doing it. The high rate of false positives is only fodder for needless fearmongering."*

- *"Mass testing of the population was not introduced to track cases of Coronavirus but to induce fear to control the masses."*

- *Asymptomatic cases defy common sense."*

- *"If they're not symptomatic, they're not going to infect you."*

- *"The Covid-19 restrictions that were introduced at the start of the pandemic (and continue to be used) don't work. Masks are incapable of preventing the spread of the alleged virus, and the lockdowns never slowed transmission."*

- *"You don't need to be vaccinated by inadequately-tested and dangerous gene-based, spike proteins, and you should ignore what corrupt scientists tell you to do."*

- *"You need to be symptomatic in order to be infectious."*

- *"What we do, we quarantine the sick. We've always done that."*

- *"We've quarantined the sick because that's how you avoid infecting the wider population."*

- *"The idea of quarantining the well with these so-called lockdowns is a new invention, and it has no foundations whatsoever either in science or in the history of controlling epidemics."*

- *"Covid-19 is less of a threat than influenza. The mortality rate is so low and the illness is clearly not causing excess deaths."*

- *"There are no excess deaths. The same number of people on average have died in 2020 as previous years. This wouldn't be the case if we really had any lethal pandemic."*

- *"If people don't wake up in the next few weeks, they will have lost the chance to take back freedoms and return to normal society as vaccine passports will be introduced."*

- *"Government policy – even before the start of the Covid pandemic – has turned decades of understanding of how to protect people from infectious diseases on its head."*

- *"Governments around the world continue to order the masses to run away and hide in their homes, being forced into unnecessary lockdowns and having to submit to tyrannical restrictions."*

- *"Excess deaths that are being contributed to Covid are mainly due to heart disease, stroke and cancer which suggests that they are actually due to the absence of medical care."*

- *"I'm warning you that governments around the world, and certainly yours locally, are lying to you in various ways that are easy for you to establish."*

- *"Don't say you weren't warned because I've been warning people as long as I can and as hard as I can, that you can still right now take your normal society back and take it back tomorrow."*

- *"I am sincere about my words and would not lie, because I am receiving absolutely nothing except criticism and social isolation from my peers."* [xvi]

Upon this somewhat provocative basis – or at least, I hope that Dr Yeadon's comments provoke some serious reflections amongst us – and, having read the reference materials quoted in the bibliography at the rear of the book, but also in the understanding that most people will not have the time or patience to wade through everything listed there even if we are in fact dealing with life-changing and world-changing events; I would like to recommend the following books as being, in my opinion, outstanding sources for the true facts that contain solid, credible explanations surrounding the Covid-19 phenomenon; the reading of which could very well mean the difference between life and death – and between freedom or tyranny.

- "CORONA: False Alarm? Facts & Figures." *By Dr Karina Reiss and Dr Sucharit Bhakdi.*
- "THE TRUTH ABOUT COVID-19: Exposing the Great Reset, Lockdowns, Vaccine Passports, and the New Normal*." By Dr Joseph Mercola and Ronnie Cummins.*
- "ANYONE WHO TELLS YOU THAT VACCINES ARE SAFE AND EFECCTIVE IS LYING." *By Dr Vernon Coleman.*
- "MY AWAKENING AND THE COVID-19 FRAUD." *by Ciaran Boyle.*
- "COVID-19: THE GREAT RESET." *By Klaus Schwab and Thierry Malleret.*

As the reader can see, we have included Dr Schwab's book as a reference as well, because it is equally important to understand the 'vision' of these globalist elites as they presume to 'reset' the world according to the requirements of the evolving establishment who are by-and-large driving the whole Covid-19 phenomenon according to their own far-reaching, and future-shaping plans. In light of new facts and data in Dr David E Martin's report, "The Fauchi Covid-19 Dossier"[7] that now answers so many of the divisive questions that have plagued us since the outset of the Covid-19 phenomenon, we will open Part Two with some clarifying paragraphs on certain key aspects of the crisis so as to bring everyone up to speed so-to-speak, before moving on to the somewhat horrifying conclusions as to what

---

[7] Also now in book & video format with the title, "PLANDEMIC INDOCTORNATION"

exactly has been going on here, and what we will each have to do if we are to survive what may yet prove to be the single biggest mortal calamity to befall humankind – as well being the single most focused opportunity to date, for a true, collective, moral-and-spiritual great awakening.

CORONAVIRUSES – the official data.
First, let's clear up a few basic facts about coronaviruses so that we are all clear about what it is that we are dealing with here. Don't worry, we're not going to get too technical, but if you were like me before we started researching then the following information should help as a base. Now, the first thing to know is that there are millions of viruses in the ecosystem, and that coronaviruses are just one family of viruses which were first discovered in the 1930's in chickens, and then in humans in the 1960's. Coronaviruses are 'zoonotic' meaning that they originate in animals and can be transmitted to humans, but only seven of the currently-listed coronaviruses are known to affect humans as listed below. 'HCoV' stands for human coronavirus.

- **HCoV-229E:** discovered in 1965 in Chicago. A common cold with generally mild symptoms.
- **HCoV-OC43:** discovered in 1967, as documented in outbreaks since 2003, also a common cold with mild symptoms.
- **SARS-CoV-1:** Severe Acute Respiratory Syndrome coronavirus detected in Asia in late 2002. Killed 774 of 8,098 sick people.
- **HCoV-NL63:** discovered in 2004 in the Netherlands. Also generally mild symptoms.
- **HCoV-HKU1:** discovered in Hong Kong in 2005. Also generally mild symptoms.
- **MERS-CoV:** Middle East Respiratory Syndrome-related coronavirus occurred in 2012 and 2018 in Saudi Arabia and in South Korea in 2015. Also potentially dangerous to some.
- **SARS-CoV-2:** Severe Acute Respiratory Syndrome coronavirus 2 – the truth about which we hope to clarify in this book.

As an added reference, and for those who may not understand the fundamentals of viruses as pathogens in general, here is a brief summary which should help us get a grasp on some of those essential facts. (Don't worry, this will be the only bit of scientific gobbledegook – and we'll try to make it as clear as we can).

"VIRUS" *(noun)*
  An infectious agent of small size and simple composition that can multiply only in living cells of animals, plants, or bacteria. Viruses possess unique infective properties and thus often cause disease in host organisms.

The various dictionary entries define 'pathogens' (or infectious, or disease-causing agents such as viruses) as biological entities that cause disease or illness to their hosts. Scientists have classified hundreds of thousands of known viruses since 1892, but say that millions more co-exist within the ecosystem. Many are symbiotic in nature and deliver crucial benefits to their hosts including some that are essential to human existence, making up about 8% of our DNA.[xvii] These are the ones that are good for us. But other viruses cause mild-to-moderate illnesses such as the common cold, chicken pox, shingles and herpes, while others are responsible for hepatitis, MERS and different types of flu – including the Asian bird-flu (H5N1) which carries a 70% mortality rate. Influenza (the seasonal flu) has been around since the 1500's and comes in types 'A' & 'B'. It is caused by the influenza virus which is similar to – but not the same as – a coronavirus. Influenza typically kills around half-a-million people worldwide each year. Certain debilitating diseases such as polio, measles, mumps, HIV/AIDS and smallpox have been largely contained or eradicated due to effective vaccination programs or anti-viral drugs, but others remain very dangerous to humans including the Marburg, Ebola, Rabies and Hanta viruses, which have a mortality rate of up to 90%.[xviii] The higher the mortality rate – the more 'pathogenic' any particular disease is said to be. SARS (severe acute respiratory syndrome) first emerged in the public sphere in 2002-2004, caused by the coronavirus 'SARS-CoV' (No.1) that mainly affected 25-70 year olds. The SARS (No.1) mortality rate was estimated – and eventually confirmed by the WHO – at around 12% of all known, proven cases.[xix]

To be more specific (because this will come into play later on) according to the WHO, a total of 8,098 people worldwide became sick with SARS-CoV (No.1) during the 2003 outbreak, and of these, 774 died. You may note that we are drawing these quotes from a 2003 article because quite frankly, any official data that has emerged post Covid-19 or even after the preparation period for Covid-19 which began with the release of the 'Lockstep' Rockefeller Report in-and-around 2010, must be considered suspect. Speaking of SARS-CoV (No.1) the WHO said: *"The fatality ratio is less than 1% for people younger than 25, 6% for those aged 25 to 44, 15% for those aged 45 to 64, and more than 50% for people 65 or older."* [xx] This averages out overall to about 12% of all known, proven cases of SARS-CoV-1 infection and is important because it is largely upon the concept of 'SARS' – as opposed to any of the other milder coronaviruses listed – or indeed of MERS – that the 2020 pandemic panic was launched.

Official sources tell us that every year, infectious disease epidemics ravage local populations, and viral pandemics such as influenza circle the globe – something we have been living with, and coping with for decades. The US Centre for Disease Control (CDC) states that up to 1 billion people contract the flu each year, with average mortalities ranging from 291,000 to 646,000, or 0.29% to 0.64% morbidity rates.[xxi] Of the four major viral pandemics of the past 100 years, the Spanish Flu of 1918 killed approximately 50-100 million;[8] the 1957 Asian Flu killed 1.1 million; the 1968 Hong Kong Flu killed 1 million; and the 2009 Swine Flu killed approximately 18,500 whilst shockingly, over a 40-year period HIV/AIDS has killed some 32 million people, or around 42% of those infected. An enormous number of

---

[8] The bacteria that caused the 'Spanish Flu' arose out of experimental vaccines given to the US Military by the Rockefeller Foundation. It is unknown if the consequences were preplanned or incidental.

previously-unknown viruses, such as the one that caused the 2003 epidemic of SARS, have claimed the lives of people and animals around the world. But interestingly, NO cases of SARS-CoV-1 have been reported worldwide since 2004 – an intriguing fact that will come into play as we progress. In just seven years (from 2011 to 2018) the World Health Organisation (WHO) claims to have done battle with 1,483 epidemics. Epidemics are outbreaks that remain confined to one country or region. Pandemics are global outbreaks.

Prospects for even deadlier, airborne microbes have risen since 1989 because the technology to alter viral and bacterial genes is now fast, easy, cheap, and precise. Whether this is achieved via CRISPR (a process to alter bacterial DNA) or even newer genetic manipulations, it is now possible to give microbes all sorts of attributes—or even to make them from scratch, from the DNA up. In light of the ban on biological and chemical weapons research, this fact will come into play shortly, particularly in respect that up to around 1999, all coronavirus-related studies and the resulting patents on research and testing procedures was exclusively confined to animals. This is because coronaviruses are 'zoonotic' meaning they are first transmitted (or so we are told) *from* animals *to* people. Unfortunately, some of the people we are trusting to inform us of the truth and the facts are heavily involved in dodgy coronavirus research and have a clear vested interest in fogging the origins of something that we now know is actually man-made – or at least is man-manipulated. And whether a man-made killer pathogen leaks accidentally, or is deliberately spread by malevolent individuals, no nation has the organization and technology to halt an outbreak once the germs escape their lab confines.[xxii] An influenza pandemic akin to the 1918 flu would today cost the world economy $3 trillion, or up to 4.8 percent of global gross domestic product (GDP).[xxiii] Why we must always put a monetary value on things invaluable – such as human life – or indeed how we presume to do so in the first place, remains beyond me.

This brings us to another very recent development; the publication of a US Congressional Report that suggests there has been an international cover-up of the true origins of the virus, and that it DID in fact emerge, "possibly as an accidental leak" out of the Wuhan Institute of Virology – but that it did so earlier than was thought, possibly in September 2019 when, surprise, surprise, the WIV did three unusual things within days of each other:

- They removed their viral database from the WIV website.

- They fielded invitations for a revamp of their lab security.

- They advertised a $600+ million dollar contract to overhaul their air-conditioning system.

In an entertaining and highly-informative video British comedian-turned-social-activist Russell Brand explains that the Congressional Report notes otherwise-unexplained increases of activity around the WIV at the time.[xxiv] The Report tentatively concludes that the Chinese Communist Party (the Chinese Government in other words) is neck-deep in efforts to fog the Covid origins story. But why? That's the million-dollar question of course. But before we leap to a eureka moment and point the accusatory finger at the CCP we need also consider what other parties may have been involved not only in the creation and release of some artificial pathogen, but also in whatever cover-ups are going on? Because as we will see as we progress through the following facts, figures and falsehoods, a number of shadowy hands in locations other than China would have every reason to want to point the finger of suspicion at the CCP so as to avoid the spotlight falling on them, and some of them are on very close and personal terms with the American body politic.

CHAPTER ELEVEN

THE COVID-19 DICTIONARY

Now, in order to make any real sense of the Covid-19 phenomenon and before we examine the truly dramatic supporting evidence, it would be helpful to all of us to understand what specific terms actually meant when they were being used by the authorities. Because as we shall soon see, they did not necessarily mean what we *thought* they meant. It is now clear that many of these scientific and medical terms were deliberately deployed in a colossal act of mass deception and psychological social conditioning, by people who knew *exactly* what they were doing!

The terms "SARS-CoV-2" and "Covid-19". Although abundantly used to describe this alleged 'novel' virus and the disease it supposedly causes, these terms could NOT (at first) be taken as absolutes that referred to something *definitively* different to existing coronaviruses and the diseases they caused. They *might* – but equally, they might not! This ambiguity was largely due to three factors: (i) the failure of the scientific establishment to *properly and definitively* 'isolate' the virus (more on this shortly); (ii) the use of the much-maligned generic **PCR Test** to identify *any* coronavirus-related debris which were then misleadingly recorded as definitive 'Covid-19 cases'; and (iii) because, (as the scientific experts who were involved in the spread of this disinformation were fully aware of) they deliberately used existing 'SARS' and 'deadly virus' terminology to mislead the public into believing that the unfolding Covid-19 phenomenon was something that it most definitely was not! (More on this too, shortly).

Interestingly, in a discussion in the U.S. Congress in May 2021, Dr Anthony Fauci the head of the Centre for Disease Control (CDC) in the United States, was specifically questioned as to his official stance that, "SARS-CoV-2 originated naturally" in light of the results of an

Australian scientific study published by Cornell University which all-but eliminates that possibility in view of the fact that SARS-CoV-2 is, "artificially engineered to bind to human cells more efficiently than to any other species". Apparently, this could NOT have occurred naturally and *had* to have been artificially manipulated. On the other hand however, and in contrast to the opinions of those who believe that SARS-Cov-2 as a medical reality is a total fabrication; that it is merely the name given to all existing coronaviruses including the common cold – plus the seasonal flu – now conveniently rebranded for the sinister purposes of the New World Order; this Australian study suggests that SARS-CoV-2 is in fact a discrete entity (however it may have originated) and that it is indeed uniquely different—albeit still very closely related—to existing coronaviruses. This is not to say however that it is in fact SARS-CoV-2 that is causing all of these reported 'cases' of something that they are calling "Covid-19" (as we shall soon see) – nor does the proven existence of SARS-CoV-2 as a separate virus – either theoretically or as a physical reality – eliminate the possibility and indeed the high likelihood that most, if not all of the supposed 'facts & data' contained in the official narrative are false, misleading, grossly exaggerated, and/or outright lies, but it helps us to really tackle the Covid-19 phenomenon in Part Two of this book if we begin at a point of common acceptance that 'something different' (which *they* are naming SARS-Cov-2) does in fact exist, although it may yet prove to be the case that its actual existence does not necessarily physically link it to the Covid-19 phenomenon, other than largely in theory, and by way of propagandic consequence.

Now I know that last paragraph could cause some bewilderment and confusion, so let me try to put it another way. The 'bad guys' needed *something* to trigger off all of these draconian social-control measures. In the absence of a real, naturally-occurring pathogen arriving unexpectedly on the scene that would, naturally, put a lot of the bad guys at mortal risk as well (which clearly hasn't happened) they have instead cooked up and literally 'created' a suitable pathogen which is allegedly 'highly infectious' and with SARS-like properties, so as to qualify it as a likely pandemic-causing agent – IF it were at large in the population. Having since created this thing through illegal bioweapons research, and/or (and this is very important) having only *theoretically* created it on paper sufficient to be able to describe its microscopic properties in detail, the bad guys only have to declare its release into the population – with enough scientific jargon and detail accompanying – to provoke all of these Covid-19 related control measures. Then, they can just label all of the normal cold and flu cases as 'Covid-19' in the hope and expectation that a full year of these supposed 'Covid cases' will be enough to get the global population under control before we all work out that the seasonal flu has apparently disappeared without trace or explanation, and that all of the supposedly 'fatal' cases of Covid-19 were of people who would have either died naturally from existing conditions, or who would have succumbed anyway to the cold and flu season.

The beauty of this scenario (for the bad guys) and something that explains Golfgate, RTE parties and off-camera political gatherings, is that the Cabal knew all along that there was no risk from any purported virus, but that there WAS a risk from the PCR tests and the vaccines – as we shall soon see. They of course, will tell us that they have all been tested and vaccinated the same as the peasantry, but really? Why on earth would we believe them? The plan it seems, was to get the agenda going by way of a declared pandemic, and then let that Covid-19 snowball gather momentum-and-power based largely upon the trust, gullibility, fear and compliance of the masses – even as we all continued to research and debate what's really going on. The pandemic was given legs in other words, not upon any real threat from SARS-Cov-2 – whether actual, theoretical or not – but through the response of the masses to 'instructions from above' from utterly untrustworthy sources. The scary part is that it is possible that they only needed a certain percentage of the population – other than those employed by the State – to comply with the Covid directions to achieve the overall goal of full effective control of the population, and in dread that this may already have been accomplished, we can only hope and pray that we are not too late in getting this message out.

In any event, given the evidence contained in Dr Martin's 'Covid-19 Dossier' that patents were filed over several years pertaining to 'gain of function' research into coronaviruses (more on this shortly), it seems that any doubt as to the existence of SARS-Cov-2 as a discrete entity (theoretical or actual) can now be parked – at least for the time being. That is not to say that further evidence of orchestrated mass deceptions may not still surface, and that sincere questioners (including myself) had every right and cause to suspect Covid-19 to be a sinister invention of the globo-political-pharmaceutical-biotech establishment; as has since been proven to be the case. Nor does this preclude the possibility that the other coronaviruses that affect humans could also be the by-product of Frankensteinian developments arising out of profit-generating bioweapons research. But for the sake of clarity and expediency, if we advance from here on the understanding that there is in fact another discrete coronavirus other than those we already knew about; one we now call 'SARS-Cov-2' which carries certain identifiable properties that differentiate it – albeit slightly – from the others. But as we will soon see, it *still* isn't what they have been telling us it is.

Adding weight to the hypothesis that SARS-Cov-2 did NOT occur naturally in another species before being transmitted to humans for example, is the fact that it only took 4 months for scientists to identify the host animal source for SARS-1 in 2003, and 9 months in the case of MERS in 2012 but to date, some 20 months after first reporting, we still haven't been able to identify a host animal source for SARS-CoV-2? Why ever not? We now know that it did NOT originate in 'another species' not unless we are acknowledging that the diabolicals who planned and created SARS-Cov-2 in the lab are indeed something less – or perhaps something essentially different – to us normal humans? We will shortly explore the evidence that shows how and why 'SARS-Cov-2' is neither novel (new) nor natural, and how and why we have been systematically lied to by the establishment in respect of almost ALL of the public data pertaining to the Covid-19 phenomenon.

"Pandemic" is the term being used by the establishment – and the World Health Organisation in particular – to get our attention. But the problem is that 'pandemic' doesn't mean today what it meant a few years ago. In context of the grasping activities of Big Pharma it is very important to note that contracts were signed-and-sealed with national governments that made them legally obliged under those contracts to order millions and billions of vaccines from Big Pharma in the event that the WHO officially declared a pandemic. Not to be too cynical here, but the facts speak for themselves. Up to 2008, the WHO definition of 'pandemic' read:

> *"An influenza pandemic occurs when a new influenza virus appears against which the human population has no immunity, resulting in several, simultaneous epidemics worldwide with enormous numbers of deaths and illness."*

This was changed in the month leading up to the 2009 Swine Flu to:

> *"An influenza pandemic may occur when a new influenza virus appears against which the human population has no immunity."*
> xxv

The omission of the, *"enormous numbers of deaths and illness"* clause allowed the WHO to conveniently declare 'a pandemic' after only 144 people had died from the Swine Flu worldwide (or so we were told) prompting the triggering of all of those legally-binding lucrative contracts; and it's also why COVID-19 is still being promoted as a pandemic today even though plenty of data indicates that the lethality of COVID-19 is more-or-less on par with the seasonal flu. The added possibility that both SARS (No 1) in 2003 and the 2009 Swine Flu could, possibly have been deliberate releases for the express purposes of qualifying the ordering of millions of doses of vaccines – by contract or otherwise – must be seriously considered, as must the possibility that the Covid-19 phenomenon comprises a similar act of preplanned global extortion. The very preexistence of private patents on those 'novel' coronaviruses itself, speaks volumes.

The plain fact is that Covid-19 was declared a pandemic by the WHO on March 11th 2020, at a point where it was claimed that 120,000 people were infected worldwide (so we were told) resulting in less than 4,500 deaths. At the time of that announcement this equalled a morbidity rate of approximately 3.75%. In contrast, the previous flu season (which was NOT declared a pandemic) took approximately 450,000 lives – a stark 100 times more deaths than those of Covid-19 even though the morbidity rate (for the flu) was only around 0.5% of those infected. This leaves us asking why wasn't the previous season's flu declared a pandemic by the WHO? And why, with such little data and facts and so few deaths, was Covid-19 suddenly such a threat?

"Cases" in normal parlance usually means, "identifiable incidents where the disease is present" and it is largely upon this definition that the official figures have been compiled. Unfortunately, many such 'cases' were no more than positive results from a PCR Test which even the authorities have acknowledged (now, long after-the-fact) to have been around 97% inaccurate. In other words, the PCR Test results that inflamed the case figures and caused massive public panic were simply NOT to be trusted because in short, these tests were picking up 'evidence' of other common viruses-including 'background noise' (the scar tissue so-to-speak) left over from previous viral infections including long-since inactive bouts of the common cold or flu. When we add in the fact – as asserted by the inventor of the PCR Tests himself, Dr Kerry Mullis – that the PCR Test was NOT designed as a viral detector and that it could easily be abused as a magnifier to purportedly 'identify' supposed evidence of any viral or bacterial infection; then we are left asking the hugely-pregnant question as to why indeed the authorities were insisting on using the PCR Test as the global go-to default testing procedure in the first place? In hindsight, it appears clear that the objective was to raise alarming numbers of purported 'cases' by any pseudo-scientific means possible. The fact that a goat and a paw-paw fruit tested positive in Tanzania says it all.

"Asymptomatic Cases". Another oxymoronic term used by the authorities to convince us that even if we are not feeling or displaying any of the listed symptoms, that we *could* nevertheless be carrying and spreading 'the disease', and therefore *must* be registered as 'cases'. It also suggests of course that *anyone* could be carrying and transmitting the disease regardless of the presence or total absence of any visible or physical symptoms. In other words, that any of us could be carrying and transmitting an invisible disease that had NO physical symptoms or side effects, and whose presence could only be detected by a curiously-unsuitable establishment-controlled test; providing a very convenient and alarmist basis upon which to lock down the whole world.

"High Rates of Infection". Another ambiguous declaration that needs to be qualified by some explanation of how many people were being tested at any given time, and in what particular circumstances? ..without which knowledge, statements like this make little sense other than in context of spreading general fear and alarm. Notwithstanding the total unsuitability of the PCR Test, the only reason for quoting alleged 'high rates of infection' would be in context of the mortal dangers posed by that infection – right? Mortal dangers that simply haven't resulted in any more deaths since the arrival of Covid-19 than in any normal year. In other words, statistical mortality isn't actually an issue here, and it should never even have entered the equation. It was tacked on to the narrative to drive the fear and the panic, and to foster compliance and obedience in the general public. Accordingly, it seems we were simply talking about a 'highly infectious' common cold at the (medical-and-theoretical) very worst, or with 'something else' such as deliberate misinformation and unfounded statistics that were being purposefully masked by pseudo-science and then publicised broadly for maximum alarmist effect.

"Vaccination" as a medical procedure is being misleadingly and disingenuously used to describe an injection process that delivers 'stuff' into the human body that; (i) has NOT been properly or adequately tested or proven to protect against a possibly-non-existent 'novel disease'; (ii) where even the authorities concede that these injectibles do not prevent transmission of the said disease; and (iii) do not prevent any re-catching of the Covid disease nor provide protection from any of its reported new variants either? This leaves the obvious question; then what exactly is 'the vaccination' and why are people being injected with it? (More on this crucial topic shortly).

"The Vaccines are 90% (or 100%) Effective!" Perhaps they are. But the real question is; effective at what exactly? (See above).

"Deadly Virus" usually means 'a virus that is deadly'. But apparently, this one (whatever it actually is) even by the official figures has an overall recovery rate of between 98.7% and 99.8%. (Some sources place it as high as 99.98%). It has proven deadly for some who were already very old or very ill – a tragedy indeed – but perhaps the term 'typical coronavirus' rather than 'deadly virus' should have been used by the authorities if accuracy was important? But of course, if they were in the habit of telling the truth, and if the truth about this purported 'global crisis' had been shared with the public from the outset, then there wouldn't have been any real crisis at all in the first place – would there?

"Spike Proteins" are the sticky-out bits that protrude from the surface of a coronavirus that serve to dock or attach themselves to the host's cells, thereby causing infection. They are important to our study of the Covid phenomenon inasmuch as they can also be en-gineered in the lab to carry out other invasive or 'passenger' functions before being injected into a host in the form of a vaccine.

"Covid-Related Deaths". This term has been greatly misused to give the impression that 'x' amount of people died directly *from* 'the virus' when even some of the more credible fact-checkers now admit that this terminology has been misapplied to anyone at all who can be remotely linked to 'the virus' even by distant association,[xxvi] even if they had personally tested negative before they died.[xxvii] The classic example of the poor soul who fell of a ladder and broke his neck only then to be classed as a 'Covid death' is a case in point, as were all of those terminally ill people who were going to succumb anyway to any unfortunate encounter with the common cold or seasonal flu. The added fact that 'incentives' (otherwise known as bribes or inducements) were being paid to numerous medical professionals by various elements of the political establishment and by Big Pharma to both; (i) push the Covid-19 narrative incessantly, and (ii) to report infections and deaths accordingly, is a shocking indictment of certain medical professionals' total disrespect for the Hippocratic Oath to, "First. Do no harm!" One can have some sympathy for those who have complied with questionable directions from above out of fear of losing their practice, but again, it all comes back to the, "just following orders" mantra – a defence that went out at the Nuremberg trials.

"Hospitals Are Overwhelmed". Only true if one concedes that they were overwhelmed with lies and exaggerations; with newly drafted-in security personnel; with creating special Covid Wards complete with unnecessary ventilators that killed far more than they saved; with stupid and inane counterproductive 'health restrictions'; with cancelled surgical appointments; with MSM crews filming hospital car parks chock-full of ambulances that had melted away less than an hour later; with staff performing 'Jerusalema' on social media; and with gagging orders that prevented (most of) the staff from speaking publicly about 'hospital business' under pain of dismissal. (Thankfully there were a few whistleblowers with the courage to speak out).

"Protect Others and Protect Yourself". I hope that everyone remembers the official push for us all to get the 'Close Contacts' app on our smartphones so that the government could advise us when we needed to self-isolate because, by happenstance apparently, we had (according to them) somehow come into dangerous contact with someone who was 'infected'. Now before we dispense with this nonsense appropriately let us first of all remember that whatever Covid-19 might be (either theoretical or otherwise) it is certainly no different and no worse than a typical cold or flu. So why all the panic and alarm and the sudden unprecedented need for all of us to be digitally tracked, traced and isolated? Again, something doesn't quite add up here. If there was a genuine bubonic plague or such like, then the notion of a track-and-trace function on our phones would be enormously helpful in preventing the spread, but a common cold or flu that doesn't even show any symptoms? No. It's just not credible.

Similarly with the instructions to 'social distance' to a distance of 2 metres, which rounds up interestingly, to 6 feet 6 inches; as well as the instructions that we could not attend weddings, funerals or social functions due to the enormous dangers posed. Well, some very clever psychologist somewhere who will probably never be named, has quite correctly informed the Cabal that if they want to manipulate a population without reverting to overt threats and physical violence, then the best way to achieve this is by fragmenting social cohesion through fear, isolation, depression and confusion, and through mental distress at not being able to even bury one's loved ones. There may also be other less-obvious but more suspicious reasons for the powers-that-be to want us all to remain a certain distance apart, such as not being able to discuss in large groups or at public gatherings the curious ability of a microscopic virus to be able to differentiate between different social locations; between different classes and ages of people; and indeed between people who paid the required €9 for a pie with their pint, and those who did not. Because apparently, and quite illogically, beer drinkers were somehow more susceptible to the virus.

Or, maybe it was because people with a few pints inside them are sometimes inclined to talk too much?

CHAPTER TWELVE

"THE VIRUS HAS BEEN ISOLATED"

OR HAS IT - REALLY?

In the interests of complete transparency, we have to admit that despite having thoroughly researched this issue over many months, that a serious question-mark remains over what exactly the term 'isolates' means in scientific terms. Because the debate has been hot and heavy, with defenders of the Covid narrative (including several high-profile MSM outlets) vehemently insisting that the virus HAS been isolated – indeed "many times!" – thus qualifying the need for, and the rush to produce experimental vaccines. Whilst independent others are equally adamant that it has NOT been *properly* isolated – thus raising questions of course about all of the supposed 'solutions' (facemasks, tests, lockdowns, vaccines) to an unknown 'something' that the authorities cannot even produce in isolation? This question is fundamental to many aspects of the Covid-19 phenomenon, because in any 'normal' set of circumstances whereby a new pathogen emerges, there are a number of ways of identifying it for the primary purpose of producing a vaccine or a cure. So, we can either look at the term 'isolates' in its literary form, meaning; "To separate from its surroundings" or, we can pose the specific question as to whether SARS-CoV-2 has been 'isolated', "according to Koch's postulates"?

Koch's Postulates are four controlled procedures usually used by the scientific establishment (since 1884) to identify the cause of an infectious disease, and without which process and the knowledge derived (we are told) it would be nonsensical to even begin to try to create an effective vaccine. Koch's Postulates were modified in 1937 by the introduction of River's Postulates which further establish the role of a specific virus as the cause of a specific disease. This was the method (we are told) by which the medical establishment determined what exactly SARS-CoV-1 was in 2003 before they began vaccine research. (Please look it up if you need the technical data).[xxviii] This is not to say that we should believe everything they are telling us, but the important point to note is that for some bizarre reason that is being obfuscated and obscured, that whichever set of postulates is used, SARS-CoV-2 fails both tests. How and why it does so will become clear as we progress, but we can give a little spoiler here by simply pointing out that Koch's and River's postulates are only designed to work on *natural* phenomena.[xxix]

Again, not wanting to get too embroiled in the scientific jargon, it appears that various piecemeal attempts to separate and identify 'the virus' were undertaken by individuals and in laboratories, but these procedures were being inaccurately described as producing 'isolates' by persons who knew full well that anyone with a medical background would naturally presume that the Koch's Postulates process had been properly adhered to.

Inasmuch as this was the case, then it was a clear deception and we should of course be asking why on earth anyone would want to mislead not only the public, but also the medical community on this crucial issue? So, has the virus been isolated or has it not? Well, lets' start by hearing what the China Novel Coronavirus Investigating and Research Team (CNCIRT) scientists have to say about, after all, the Wuhan Institute of Virology is probably *the* best placed location to begin our search for answers.

Their study was published in the New England Journal of Medicine on January 24th 2020. It was headed, *"A Novel Coronavirus from Patients with Pneumonia in China, 2019"* and explained how the Chinese scientists had extracted RNA samples from infected lung tissue to effectively create a *model* of SARS-CoV-2, which they then shared with the WHO for "surveillance and detection" and to "refine strategies to prevent, control and stop the spread of 2019-nCoV" (the original name of SARS-CoV-2), saying:

"Further development of accurate and rapid methods to identify unknown respiratory pathogens is still needed." (and) *"Although*

our study does not fulfill Koch's postulates, our analyses provide evidence implicating 2019-nCoV in the Wuhan outbreak.[xxx]

Mmm, so for some reason, the Chinese scientists (and a great many of them too) did not use the standard Koch's postulates in what should have been a relatively straightforward testing procedure with a pure sample of the virus at hand? Compounding the matter greatly, but at the same time confirming the great lie are the following two quotes from the USA's Centres for Disease Control & Prevention ('CDC') which simply do NOT add up. Noting the fact that this predates the CNCIRT report by 4 days, the first quote attempts to provide a timeline for the isolation, production and distribution of SARS-CoV-2.

- *On January 20, 2020, CDC received a clinical specimen collected from the first reported U.S. patient infected with SARS-CoV-2. CDC immediately placed the specimen into cell culture to grow a sufficient amount of virus for study.*

- *On February 2, 2020, CDC generated enough SARS-CoV-2 grown in cell culture to distribute to medical and scientific researchers.*

- *On February 4, 2020, CDC shipped SARS-CoV-2 to the BEI Resources Repository.*

Well, given that this information is on the CDC's SARS-CoV-2 Viral Culturing webpage as of July 2021 complete with that dated timeline that was updated in December 2020, we then need to find an explanation – other than outright lying – for the following extract from an internal CDC Report of July 2020 that unambiguously says;

"Since no quantified virus isolates of the 2019-nCoV are currently available, assays [diagnostic tests] designed for detection of the 2019-nCoV RNA were tested with characterized stocks of in vitro transcribed full length RNA." [xxxi]

In simpler terms, *"We don't have a live sample of SARS-CoV-2 (but please don't ask us why not) so we have made some up out of similar coronavirus stuff that we already have in the lab so that you all know what to look for. Oh, and very coincidentally, the PCR Tests are equally sensitive to any and all debris from that similar coronavirus 'stuff' that we are using to create a supposed gold-standard 'isolate' of SARS-CoV-2. Or at least (shhh), ...that's what we're telling everyone."*

To add to the intrigue, there's this response from a Freedom of Information Request to the UK's NHS, who in turn referred the applicants back to the European Centre for Disease Prevention and Control (ECDPC) whose Eurosurveillance Journal unambiguously confirms that; (i) they are using 'respiratory samples' (not isolates) and that the PCR Tests do NOT distinguish between live viruses and dead viral debris:

> *"Virus detection by reverse transcription-PCR (RT-PCR) from respiratory samples is widely used to diagnose and monitor SARS-CoV-2 infection and, increasingly, to infer infectivity of an individual. However, RT-PCR does not distinguish between infectious and non-infectious virus. Propagating virus from clinical samples confirms the presence of infectious virus but is not widely available (and) requires biosafety level 3 facilities".[xxxii]*

So, it appears that the CDC, the NHS, the WIV, the ECDPC and the CNCIRT and all of their affiliate institutions didn't actually have any virus isolates at the time, despite the CDC's latter claims to the contrary? But why the lies, and how and why didn't any of these agencies raise the red flag about these inconsistencies?

Now, I know this seems like a lot of scientific gobbledegook, but there are two very, very important pieces of information here which, in simple layman's terms means; (i) that even the CDC, as late as July 2020 and some nine months into the pandemic – and despite their latter claims to the contrary – did NOT have a live sample of the new virus. And (ii) that they were all, somewhat unbelievably, qualifying the effectiveness of their diagnostic tests (to prove the existence of that speculative unavailable virus) by using pre-existing stocks of laboratory-cultured RNA. (RNA is a molecule found in humans that is very similar to DNA). In other words, they were creating tests for Covid-19 *without* a sample of the specific virus they were testing for, and then 'proving' the efficacy of those tests against 'stuff' (including an assay element called luciferase) they had just grown in the lab using a highly unreliable and inappropriate PCR Test that was going to pick up on ANY coronavirus-related debris anyway, living or dead!?

Then there's the April 2021 report headed, somewhat incredibly; *"Laboratories in US can't find Covid-19 in one of 1,500 positive tests"* where we learn that a number of clinical scientists and immunologists from seven different US Universities intend suing the CDC for 'massive fraud' after 1,500 supposedly 'positive' Covid-19 samples proved to be nothing other than influenza A or B. They discovered this interestingly, by refusing to use what one scientist described as, "the bullshit PCR Test" and instead, as per proper scientific practice, they used Koch's Postulates.[xxxiii]

Now, I don't know about everyone else, but to me, this seems both preposterous and utterly outlandish that the whole world has been plunged into this unprecedented crisis and the subsequent, now more-or-less mandatory vaccination program based upon purported 'science' like this? What on earth is going on here – because there are only three logical reasons for the CDC stating that they did NOT have a sample of this supposed killer virus in that July 2020 Report, even as they were ostensibly designing and creating tests to identify it.

(i) The first possibility is that they genuinely didn't have a sample because 'the virus' didn't actually exist as a discrete, physical entity at the time of the July 2020 Report, other than in theory. Otherwise, how do we possibly explain how the CDC – one of the largest and best-funded scientific facilities in the world, didn't have an isolated sample of a virus that was supposedly infesting the whole world?

(ii) The second possibility is that all of the scientific boffins involved at the CDC are totally inept, stupid, sycophantic morons who initiated these pointless culturing-and-testing procedures for a theoretical virus 'under orders' without asking the all-too-obvious question as to what the hell they were all playing at!?

(iii) And the third, even more disturbing possibility (because it is the most plausible one) is that the main players at the CDC and in several other such trusted organisations like the World Health Organisation ('WHO') the NHS, the ECDPC and others knew full well that it wasn't even necessary to provide physical evidence of a pathogen which, upon closer examination might then expose some sinister realities about certain key players and the roles they had in unlawful 'gain of function' research; but that it was enough to convince the public of its existence and of its mortal threat, and of course, of the urgent need for everyone to get tested via methods which were already designed to 'find' ever more evidence of this otherwise 'unavailable' killer virus, which by their own admission, they didn't even have a sample of at the time!? The latter 'timeline' claim of a convenient sequence of purported events that ostensibly puts the, "where's the isolated virus?" question to bed in the public eye, is utterly implausible unless we accept that those responsible for the July 2020 CDC Report were completely unaware that their colleagues had, some six months earlier, collected, isolated, cultured and disseminated SARS-CoV-2?

In light of very recent evidence that it was entirely in the interests of certain key players NOT to have anyone other than themselves properly identify what exactly SARS-Cov-2 comprised of – either theoretically or in substance – we can see the motivation for the deliberate sewing of confusion and misdirection as public panic took hold. Because the objective was always to cause panic and alarm – something that would have been considerably nullified had the question of what exactly SARS-Cov-2 was, had been properly answered at an early stage. But all we (public) were supposed to know was that a 'new killer virus' was on the loose, which, just like two of its six coronavirus cousins (SARS & MERS) *could* cause death. The public wasn't informed however, that despite trying since the 1950's the medical establishment hasn't been able to come up with a vaccine for the common cold (another coronavirus cousin) or indeed for any of the other five coronaviruses that affect humanity, raising the troubling question of how could Big Pharma in the shape of Pfizer, Moderna, Johnson & Johnson and Merck *et al*, all have been so quick off the mark in producing billions of purported 'vaccines' which they appear all-too-ready and eager to now test out on the trusting public? Especially in light of the fact that the average time for the production of an effective vaccine is between 7 and 20 years? Again, things just don't add up here – so... why aren't we asking more questions?

And what then of our immune systems and the notion of collective 'herd immunity'? You know, that well-established scientifically-based fact that our immune systems are more than capable of dealing with any cold or flu-like viruses that come and go as usual every year. In fact, given that the scientific community has NOT been able to produce an effective vaccine against the common cold, then it is precisely via the agency of our immune systems that the body reacts to these pesky viruses and produces antibodies that will then protect us from specific infections for life. And when enough people in the population have been infected and recovered, that's what we call 'herd immunity'. So, whilst remembering that we are still talking about reported 'cases' resulting out of flawed PCR Tests that could be picking up any sort of viral debris, how then do we align the reported fact by Forbes Magazine and The New York Times that people who allegedly 'had' the virus have now developed immunity, possibly 'for life'?

Could it possibly be that our bodies have reacted to SARS-CoV-2 (whatever it might be) in exactly the same way that we react to similar coronaviruses every year – without any need for masks, social distancing or vaccines in order to acquire individual and collective immunity? The fact that one prominent researcher advised Dr Fauchi in an email dated January 31st 2020 that the differing or 'unusual' properties of the studied virus were less that 0.1% of its genome[9] raised again, the all-too obvious question of how different any new virus needs to be in order to qualify as a new, different and separate organism other than a plain old run-of-the-mill common coronavirus?

Obviously, we need to know exactly when the scientific establishment came to an empirically-based consensus on what exactly SARS-CoV-2 constituted, and how they could possibly do so absent a 'quantified virus isolate' and in that specific knowledge we can then unravel the reasoning behind the impressive – if not alarming speed – at which Big Pharma then rode to the rescue – appropriately indemnified by our governments of course, against any penalties or lawsuits should their products be problematic in any way. Whatever the final explanation, there is no getting around the question of how and why so many people are now being coerced into taking injectibles that have NOT – by any measure of the term – been 'tested over time'!?

We will shortly examine the astounding evidence that SARS-Cov-2 was NO mystery to many of the key players involved in the official narrative inasmuch as several of them were involved in the research and creation of what is in effect a viral bio-weapon. A bio-weapon that could be equally devastating to society whether it was to be physically or even theoretically 'released' by people who had the heads-up long in advance of any likely 'accidental release' that would then require the distribution of billions of profit-making, and agenda-shaping vaccines for a virus which, even in abstract, theoretical form, could also be used as a tool of mass social control.

---

[9] Genome: An organism's complete set of genetic instructions containing all of the information needed to build that organism and allow it to grow and develop.

CHAPTER THIRTEEN

PLAYING GOD

GAIN OF FUNCTION RESEARCH

When we started writing this book we didn't know what 'gain of function research' meant. But we do now, and it is a very disturbing discovery. 'Gain of Function' in layman's terms means playing around with an organism to see what 'gains' may occur if we do this-or-that with it? Undoubtedly there is knowledge to be gained and benefits to be had from such research, but 'gain of function' in context of the Covid-19 phenomenon is understood within the scientific research community to mean, "Seeing how dangerous and infectious we can make a pathogen, such as a coronavirus." The reason this is done (we are told) is so that responses and antidotes will be ready-to-go if we suddenly get hit with something similar that emerges out of bats or pangolins; or, if 'the enemy' (such as Russia or China we suppose?) decides to use biological weapons against the USA for example. Or, that we really need to do this research so we are ready for any unanticipated accidents or spills from insecure laboratories. But unfortunately, we're not really buying these explanations. It is far more likely that some of these highly-placed scientists have gotten carried away with their own importance. That they have become mesmerized at the secretive millions in funding they can draw down from competing governments and militaries; and that they are personally excited – and quite naturally so – at the prospect of making some new discovery or breakthrough that will get their names embossed in Latin or Greek in the medical archives or patent libraries. But none of these justifications for messing with nature in such a dangerous manner are very convincing, are they, and we haven't even mentioned the diabolical possibility that gain of function research is *the* primary means – through the creation of new infectious diseases – by which Big Pharma will guarantee its own, highly lucrative future, even at the expense of the health of billions?

148

At this point, we need to travel back in time in order to put the whole Covid-19 story into its proper context; that of a series of collaborative interchanges between a broad range of players going back at least twenty years; a collaboration of insider elites that includes all of the worst elements of the amoral masculine in the reckless pursuit of power, prestige, possessions and profit, and upon which basis they intend literally, to take over the world.

Let's begin with that bombshell from Dr David Martin's Report "The Fauchi Covid-19 Dossier" which, before we go any further must be highly commended for the exacting, exhaustive and data-specific detail therein which will leave no doubt in the minds of any objective reader that the Covid-19 phenomenon is in fact the product of a number of criminal acts that have been compounded and exacerbated by the measures since taken by the authorities, to supposedly 'combat Covid's effects'. In fact, Dr Martin describes what we are calling the Covid-19 phenomenon, as "Covid terrorism". [xxxiv]

Now why would such an eminent, respected professional use such an inflammatory term? Well, one answer is because he is honest, and another is because he is not beholden to the establishment and therefore doesn't need to mind his P & Q's. There are of course numerous shady figures involved in this 'Covid terrorism' at different levels but in context of Dr Martin's report there are six main person-alities pulling the strings whose bios (very briefly) read as follows.

Dr Anthony Fauchi: Has been the head of the US National Institute of Allergy and Infectious Diseases (NIAID) since 1984 and is the Chief Medial advisor to President Joe Biden – having previously held the same role for President Trump. He has served under a number of US Presidents dating back to Ronald Regan. Fauchi is an immunologist who has done much work associated with viruses, pandemics and vaccination programs whilst serving with the National Institute of Health (NIH) and is the highest paid civil servant in America.

Dr Peter Daszak: Is a British zoologist now based in the US and a world expert on all things coronavirus-related, especially concerning the transmission of viruses between animals and humans. He is the President of EcoHealth Alliance, which is listed as, "a non-profit non-governmental organization" but which nevertheless 'administered' over $100 million in US federal grants that funded research in foreign laboratories, including the Wuhan Institute of Virology in China. Between 2014 and 2020 Daszak worked specifically on bat-related SARS-type viruses in a project funded by the NIH and Fauchi's NIAID.

Dr Ralph Baric: Is a Professor of epidemiology, microbiology and immunology at The University of North Carolina, Chapel Hill. He is heavily involved in gain of function research and contributed to the classification of SARS-CoV-2 in a paper entitled, "The Species Severe Acute Respiratory Syndrome-Related Coronavirus: classifying 2019-nCoV and naming it SARS-CoV-2" and where all the contributors declared that they had, "no competing interests".

Dr Shi Zhengli: Also referred to as 'the Chinese Batwoman' is a virologist who researches SARS-like coronaviruses at the Wuhan lab. She has worked closely with Ralph Baric and Peter Daszak and was adamant in her defence that the virus could NOT have originated at the Wuhan lab. She published a paper with Baric in 2015 describing the dangers of emerging coronavirus without mentioning the fact that any such 'emergence' would be a direct consequence of their own unlawful 'gain of function' research.

Dr Julie Gerberding: Director of the CDC from 2002 to 2009 before taking a post as President of Merck Vaccines in 2010. Ms Gerberding earns her place in this eminent line-up mainly because of her personal 'involvement' in the SARS-CoV-1 outbreak and because of some seemingly-shifty dealings in SARS-related patents at the time.

Bill Gates: Billionaire oligarch, co-founder of Microsoft who has reaped billions out of his investments into vaccines. A lifelong eugenicist, he set up the Bill & Melinda Gates Foundation in 2000 as a private charity which has since donated sizable amounts of money to various medical organizations and scientific research programs including the WHO where the B & M Gates Foundation is listed as one of the top donors, competing for top spot with the UK and the USA.

The Fauchi Covid-19 Dossier's special value is in all of the detail and cross-references and irrefutable facts that it contains which absolutely – if not incontrovertibly – point to a series of collaborative overlapping agendas by a number of the parties and institutions involved to foist a massive social hoax on the population of the world. Now this is not to say that these underhanded collaborations were all executed seamlessly or that all of the actors played their respective parts without errors or oversights, as we are seeing now for example in the FOI release of emails between Anthony Fauchi, Peter Daszak and others which provides us with the proverbial 'smoking gun' – or at least one of them – that will ultimately lead us to the perpetrators of this crime against humanity, and from there, hopefully, to an understanding as to how this diabolical fiasco could have occurred in the first place, and what we must do to prevent it happening again.

I urge anyone and everyone who has not yet read that 200 page Dossier or who has not fully digested the contents of the parallel video "INDOCTORNATION" (yes, that's the correct spelling) to please do so without delay, because it is possibly *the* single most succinct, comprehensive and convincing piece of reporting to date, that clears away most of the fog and contradictions that still plague the Covid-19 debate. Hopefully this book will also assist in this effort.[xxxv]

Planning a Global Crisis. The Dossier outlines how on October 18th 2019, some five months before the declaration of the Covid-19 pandemic, the World Economic Forum and the Bill & Melinda Gates Foundation (BMGF) sponsored a symposium in New York called 'Event 201'. It was introduced as 'a fictional event' that would simulate the meetings that would need to be held by the various stakeholders in the event of a coronavirus-like pandemic. Those playing the (supposedly fictional) roles of the Pandemic Emergency Board (PEB) included professionals from various corporations, scientific institutions, universities, the military and the political establishment with the first, very telling statement opening the proceedings that, "Governments and private corporations would have to partner together to combat the emerging threats to society." This was immediately followed by a stark warning about the predicted conspiracy theories and fake news that would abound, and of the need to swiftly and efficiently deal with anything that countered the official narrative. The Event 201/PEB symposium continued playing out a predicted, scripted narrative that aligns so exactly with the way that things have since turned out that it simply cannot be dismissed as a coincidence. Many of the key players who are currently involved in driving the Covid-19 narrative – and profiting personally and massively from it – were sitting on that fictional PEB Board.

The next section of the INDOCTORNATION video focuses upon the treatment that Dr Judy Mikovitz received at the hands of the establishment and Big Pharma when she attempted to raise questions about the whole Covid-19 phenomenon, including being targeted, threatened, discredited and vilified in the press and then eventually jailed without charge after losing her position and funding. In other words, the typical treatment that is usually meted out by the Cabal to determined whistleblowers and truth-seekers. Great credit goes to Dr Mikovitz for not succumbing to the intimidation and for declaring her intention to, "continue telling the truth" – whatever it costs!

Dr Martin then narrates the unfolding horror story surrounding the granting of US patents to some of the key players for coronavirus research since 1999, including patents for research, for testing, and for vaccinating against coronaviruses which, in contravention of the international ban on bioweapons research, big players such as Gates, Fauchi, Baric, Zhengli, Daszak and Gerberding were both funding and profiting from; apparently, with the full knowledge and support of the various institutions they were affiliated with. (Please excuse any necessary generalisations as we focus on the most important facts here).

A very important point is made about the fact that one *cannot* patent nature. It is simply not allowed. So how, we might ask, could these people patent coronaviruses that emanated *naturally* out of animals? Well, the obvious answer is because these were NOT natural corona-viruses, but were the products of artificial creations by Frankensteinian scientists who – whilst being clandestinely funded by the taxpayer – now sought to profit personally from the creation of bioweapons using gain-of-function research. But it doesn't stop there, because you can't have one without the other. If people are patenting coronaviruses which by their own definitions have been 'manipulated and altered' into something 'non-natural' then they are by definition conducting unlawful, criminal experiments into bioweapons. Full stop!

The 2003 SARS outbreak in China prompted the CDC to patent that virus too, including the disease itself and any testing procedures used to identify it, which, as Dr Martin explains, means that from 2003 to 2018 these central players, *"controlled 100% of the cash-flow that built the empire around the industrial complex of the coronavirus."* This was confirmed in a news briefing in May 2003 where Julie Gerberding said the CDC filed for a patent on its discoveries regarding the SARS coronavirus in order to protect public access to the material.

> *"We could be locked out of the opportunity to work with this virus if it's patented by someone else, and so by initiating steps to secure patent rights, we assure that we will be able to continue to make the virus and the products from the virus available in the public domain."*

But again, how could they patent a naturally-occurring disease in 2003 without admitting that it was actually 'non-natural' thereby putting the creator(s) of SARS-CoV-1 in open violation of the international treaties that have banned chemical and biological weapons research since World War I? A ban by the way that was further reinforced in 1972, and again in 1993 by prohibiting the development, production, stockpiling and transfer of these types of bioweapons?

Well, the simple answer is that the creators of SAR-CoV-1 *were* in violation of international law which means that the Covid-19 phenomenon (however we describe it) is the result – either directly or indirectly – of some serious violations of international terrorist laws. We are also left asking the question as to whether or not the release of SARS-CoV-1 in 2002-03 was deliberate or accidental, because it most certainly isn't of pure natural origins – is it? We do know that Dr Ralph Baric of UNC had been messing with coronaviruses since 1999, and had applied for patents in October 2002 for 'Methods to Produce Recombinant Coronaviruses' *before* the reported 'unexpected arrival' of SARS (No 1) in February 2003. We also know that substantial monies were then poured into the ostensible 'search for a cure' which Baric, Zhengli, Daszak, Gates, Gerberding and Fauchi would all have been involved in, but that search for a cure – along with the disease itself – then mysteriously disappeared before too long, leaving Dr Baric and his associates, with their newly-registered patents, sitting in pole position should something like this ever happen again. Is it really such a leap of the imagination now to harbour an educated guess as to SARS's *real* origins?

In May of 2005, Jim Yardley of the New York Times wrote:

> *"Not a single case of the severe acute respiratory syndrome has been reported this year [2005] or in late 2004. It is the first winter without a case since the initial outbreak in late 2002. In addition, the epidemic strain of SARS that caused at least 774 deaths worldwide by June 2003 has not been seen outside of a laboratory since then."*

Then, in 2007, during the tenure of Dr Julie Gerberding the CDC filed an application with the US Patents Office to keep their previous coronavirus patent applications secret. Really? So what happened to Ms Gerberding's previous declaration in 2003 that they were registering SARS patents so as to, *"be able to continue to make the virus and the products from the virus available in the public domain."* So... why the sudden secrecy Dr Gerberding? Had the mission changed from one 'in the public service' to a covert profit-making venture that was arguably also a criminal enterprise? Or, had you all been lying to us all along? Regardless of the impropriety of this 'secrecy' request, the fact that they had by now patented all of the elements, the research and the procedures meant that they had full and total control over all things coronavirus-related (or at least over the pathogenic bioweapons they had secretly created) meaning that no-one else could legally do anything in respect of these supposedly 'novel' coronaviruses without the express say-so of the patent holders – the CDC. Any such third-party requests risked having the secret patents flushed out of course, but who, really, other than the Patent Office or bioweapons experts would have known anything was amiss? Meanwhile, the CDC would at least be in pole position as we said, should the opportunity arise to be able to cash in on the conspiracy, and, as Dr Martin says, "turn pathogen into profit" because, to put it in Dr Gerberding's own words in a 2014 interview;

> *"People across health systems everywhere are realizing that no single organization can solve the complex challenges we face. Whether containing the current Ebola outbreak, increasing vaccination rates or improving the overall health of people in communities, **partnerships between public and private sector organizations are essential**. Merck has a unique role and a responsibility to contribute, and I welcome this opportunity to lead teams of experts who will help us achieve our mission of saving lives and improving health around the world."*

So refreshing isn't it, to hear that the Merck executives felt they had 'a responsibility to contribute' and how rewarding for them, and for Julie Gerberding in particular, when those 'responsibilities' drove

Merck – largely under the guidance of Dr Gerberding in her specially-created new role as the, 'Executive Vice President for Strategic Communications, Global Public Policy and Population Health' – to rise to the top of the vaccine-earners list? Without passing any remarks or aspersions on Dr Gerberding's personal involvement in the SARS-patenting scheme other than noting that in a May 2020 interview she stated that she had, "been involved very much in the SARS outbreak in 2003" and this was shortly after vacating her post as a tenured faculty member in the University of California's Infectious Diseases Department (UCSF). In that same busy year of 2003 she co-authored an article with Dr Anthony Fauchi in Science Magazine promoting a "global vaccine enterprise," not forgetting of course that 2003 was also the year that the CDC had applied for the SARS patents that they would later attempt to hide. Now hobnobbing regularly with the FDA, the NIH and the CDC as a Vice-President of Merck, and whilst still maintaining her role as Adjunct Associate Clinical Professor of Medicine at UCSF, Ms Gerberding epitomises the type of person who can move seamlessly from a role as a senior educator or public servant to a position with Big Business with no apparent conflict of interest, thus demonstrating that the marriage between the two, and the crossover of public and private interests between them is not nearly as separate, transparent nor as independent as we may think.

No wonder she is an advocate of 'essential partnerships' between the private and public sector, and no doubt, when the University of California San Francisco (UCSF) awarded Dr Anthony Fauchi its prestigious Merit Medal for his sterling work on vaccines and on AIDS in particular, there was no conflict of interest in them doing so either.

In 2013, under the Obama Administration, gain-of-function research was officially suspended – well sort-of, but not really - because it was in actuality only a paper exercise to give some plausible deniability should someone point out that the NIH was in effect, sanctioning illegal bioweapons research. But then, a couple of years later when questions began to be raised about the moral and ethical legitimacy of funding this highly dangerous work through Harvard University and the University of North Carolina for instance, the decision was made to offload the work abroad, specifically to China, and specifically to the Wuhan Institute of Virology – which is where Peter Daszak's EcoHealth Alliance comes in, posing as the recipient of legitimate funding whilst channelling US taxpayers money over to Wuhan.

Respected virologist Dr Meryl Nass was the first person to link anthrax outbreaks in late 70's Africa to deliberate acts of biological warfare by the Rhodesian Military, and saw troubling similarities in the Covid-19 story. She notes that a much-quoted paper written by five named scientists in defence of the 'natural origins' Covid theme simply, "did not hold water," and was both surprised and confused how it was being championed by fact-checkers and being held as gospel in the mainstream media (MSM), wondering how and why the article got so much traction other than by some sort of 'arrangement' or via 'orders from above'? Nobel Laureate Professor Luc Montaigniur also states quite unequivocally that 'the virus' was not at all natural, but was the meticulous work of, "some very expert molecular biologists".

The video goes on to demonstrate how a network of MSM and social media outlets are collaborating quite obviously, to advance the official narrative and are actively suppressing all opposing views using censorship and fact-checkers. It also shows how this propaganda is happening worldwide, and indeed, how most of us are being deceived by this misinformation and may be inadvertently spreading it too. Google comes in for some special attention because of its overarching reach and prominence and because of its highly-secret 'blacklists' whose existence Mark Zuckerberg has tried to deny, even before the US Congress. Interviews with various highly-placed officials at the CDC blatantly lying on camera about their foreknowledge of, "neurological complications" and other side effects when they ordered every American citizen to get the swine flu jab in 1976 only pours coals on the inflammatory claim that if there's profit in it – then they *will* do it – regardless of the costs or consequences on those they are doing it to! Reckless and cavalier practices that result in disabilities, sickness and death amongst the population who trust what the CDC is telling them is also highlighted, as is the reported claim that the CDC is NOT operating as a Government Agency but is in effect a privately-owned subsidiary of the pharmaceutical industry. Whether this is to be taken literally or by allusion the implication is clear; Big Pharma 'owns' the CDC, and the CDC is in the 'business' of vaccines.

Dr Martin then goes on to explain how the integrity of traditional medical journals and scientific outlets have been sorely compromised by the overwhelming wealth, power and influence being projected at them as the Cabal stacks the shelves with their own 'officially sanctioned' articles and papers that were produced by bought-and-paid-for 'experts' who endorse the official narrative – simply because they have been paid to do so. **John D Rockefeller** is featured as the world's first billionaire who apparently set the now-familiar pattern of 'controlling the narrative' via a compliant Press Corps, so as to acquire massive power, wealth and control of society. Other institutions such as the CIA are featured, including 'Operation Mockingbird' that positioned over 3,000 trained CIA operatives as managers, broad-casters, news anchors and practiced disseminators of misinformation in media outlets since the 1950's so as to control public information in order to suit 'the agenda' – whatever that agenda may be in the moment. The overarching ambition of 'the establishment' to control the individual through covert and deceptive means is something that we all need to 'get' pretty smartly, because if we are not even aware of it in the first place, then how can we possibly defend ourselves?

The Telecommunications Bill of 1996 was signed into law by President Bill Clinton. It effectively opened the door for global corporations to seize control of the media, whilst closing the proverbial door on the principles of objective truth, transparency and journalistic integrity. Examples are then shown in the video of multiple supposedly 'independent' news channels repeating the exact same news-scripts verbatim – word-for-word. Something we would not be able to notice of course unless we were watching multiple news channels simultaneously – right? But the evidence is right there for all to see as shown in this snapshot of nine different-and-independent regional US newspapers all carrying the identical story. And it could just as easily be 99 or 999:[xxxvi]

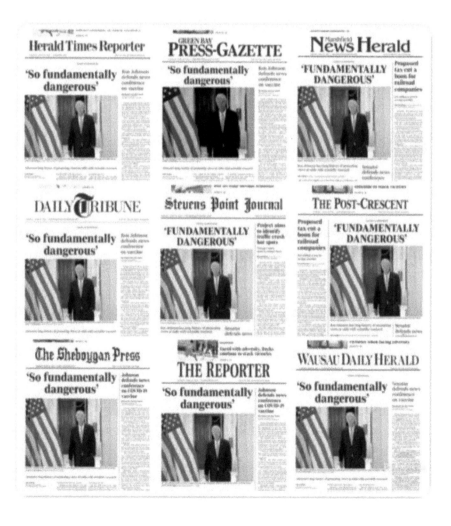

Similar examples could be taken from any large country in the world. The message we all need to be taking away from this, is that 'the media' does NOT work for us, the people. It obviously needs us to buy into the lie – literally – but it's the interests of the establishment and the orders of the Cabal that take precedence here, and we should be in NO doubt whatsoever about that! Nancy Pelosi, the leader of the US Democratic Party even explains the 'tactic' of using the MSM to smear and destroy reporters, whistleblowers or truth-seekers who challenge or defy the official narrative, and this is not something exclusive to America either. Ireland in particular is adept, if not expert at the practice, with *The Irish Times* for example, colluding with the Irish Courts to spread misinformation and disinformation in order to blacken the names and reputations of outspoken pro-justice activists and to protect from exposure, the sordid goings-on in high places.

If the reader will indulge me, two recent incidents spring to mind. The first was when National (State-sponsored) Radio called me up unexpectedly, claiming that they were doing a program on 'poison-pen letters'. I was to be one of several interviewees who had been subject to anonymous attempts at character assassination via the means of false, sexually-themed allegations, even though mine had occurred over 8 years previously and had long since been dealt with in the Courts, where, after identifying the politically-connected perpetrators, we were awarded a substantial settlement. We should have been a touch suspicious I suppose, given that the call from *RTE Liveline* came less than 48 hours after we had alerted the DPP's Office that we were applying for the Director's impeachment due to multiple proofs of serious criminal activity ongoing in those lofty chambers. Anyway, it soon became obvious that the whole purpose of the hour-long broadcast was to have the presenter read out the disgusting contents of an already proven-to-be-false report that was sent to the Child Protection Services and published online. No other interviewees joined the program and, because I held my own in the exchange and basically used the opportunity to encourage people to 'fight back' if this were to happen to them, the podcast of the live broadcast disappeared within minutes from the RTE website. Fortunately, we had the foresight to record it ourselves. You can watch the video-broadcast here, or do an online search under the title, "Attacks, Assaults & Poison-Pen Persecutions – but NO DPP Prosecutions!?" [xxxvii]

The other recent incident was when a corrupt High Court judge desperately needed to shut down a case that was becoming a massive embarrassment to the establishment; one where we had legitimately applied to the Houses of the Oireachtas (Parliament) for certain 'Officers of the Court' including that same judge, to be arrested, charged and prosecuted for a broad range of in-your-face offences against justice such as perjury, fraud, deception, forgery, conspiracy and contempt of Court. As per usual, they had no valid, lawful response to our applications, so they did what they always seem to do in such cases. They embarked on yet another conspiracy to pervert justice only this time enrolling *The Irish Times* as a co-conspirator as well. The next thing we knew, an article had been published in *The Irish Times* on May 18th 2021, where apparently, their respected and experienced Court Reporter claimed she had been present in the High Court that very morning when the judge read out a judgment that effectively shut down our case. Conveniently – albeit unlawfully – this also meant that *our* valid applications to have these criminal miscreants held to account 'under the law' were now stuck in everlasting legal limbo.

Please remember that this was OUR Court case – where we were the applicants – yet we weren't even notified by the Courts Service that this alleged High Court hearing was happening – which is yet another serious (and criminal) breach of legal procedures as anyone who has been to Court will know. Add in the fact that we now know that the purported 'hearing' never even happened at all (another recurring phenomenon in the Irish Courts) and you start to get the general picture. The Cabal will protect themselves at all costs, even through obvious, overt, in-your-face criminal acts because they have no fear of being ever held accountable. The fact that three months later the

Irish Courts are *still* refusing to provide us with a stamped-and-sealed version of that judgment document (which by law we are obviously entitled to) so as to prevent us having any documentary proof to back up a claim of criminal conspiracy between *The Irish Times*, the Courts Service and the High Court Judge in question is just more hypocritical icing on a putrid, rotten cake.[xxxviii] That judge, and 40 more of his colleagues are named in the *'Criminality in the Irish Courts'* book, which despite being a hot seller for over a year, has not yet been challenged on fact.

Returning to our synopsis of the Fauchi Covid-19 Dossier; Dr Martin goes on to explain how, in 1979, the American Government changed the patent laws so that scientists and inventors – even if they worked for the State or were directly funded by the taxpayer to the tune of millions of dollars, could register private patents on their discoveries or inventions, and then re-sell them back to the public at a profit. Likewise with the Big Pharma giants. They can take millions and billions of public money to develop a product and then – quite lawfully – set their own inflated prices on the results.

Anyone with half-a-brain or any sense of propriety must see that this simply isn't fair. If the taxpayer is funding the research, then the taxpayer should reap the benefits without having to dip into their pockets again – right? After all, these scientists and researchers are already being paid for their work and pharmaceutical giants like Pfizer and Moderna are amongst the richest corporations in the world. Here again, we see the predatory, exploitative attitudes that drive this type of ruthless profiteering and greed-mongering at the public's expense.

CHAPTER FOURTEEN

PERNICIOUS PARTNERSHIPS

PLANNING THE PANDEMIC

The Global Preparedness Monitoring Board (GPMB) is 'co-convened' (whatever that means) with the World Health Organisation and the World Bank Group. Amongst the 14 members of the GPMB Board include; a former Director General of the WHO; a former Secretary General of the International Federation of the Red Cross & Red Crescent Societies; the President of the Global Development Program of the Bill & Melinda Gates Foundation; the Director General of the Chinese Centre for Disease Control & Prevention; and various other highly-placed notables from corporations and government agencies. Remembering that the WHO is "a specialised agency of the United Nations", Dr Anthony Fauchi is now listed as a 'former' Board Member, indicating that his status at the GPMB has changed since the publication of the Fauchi Dossier.[xxxix]

Again, these global-corporate-political interconnections are noted not so much to indicate any purported wrongdoing *per se*, but simply to underscore the reality that these people are ostensibly working as an establishment-funded team and, should anything inadvertent or unfortunate or embarrassing occur, that at least they are all in a position of sufficient power, authority and influence to deal with matters promptly and efficiently. All we need to concern ourselves with here is their motives, their methods and intentions when they published a Report in September 2019 entitled, "A World At Risk" declaring the need for two global preparedness exercises before September 2020 for a possible pandemic that was triggered by a respiratory pathogen. This in turn led to Event 201 in October 2019 as discussed previously, where we see some of the GPMB Board members playing key roles. Certain uncanny 'predictions' were made at Event 201 that belie any doubts as to the foreknowledge of these big players as to what was about to unfold less than a month later, and if anyone still questions this reality, they need only do an online search of 'Event 201' or 'Clade X' to come to the inescapable conclusion that whatever Covid-19 is, it is neither 'new', unanticipated or unexpected.[xl]

Then there's the bizarre fact that Anthony Fauchi, speaking from the US President's podium in January 2017 would state that, *"There is no question that there will be a surprise* [viral] *outbreak during this administration.."* meaning *before* the end of 2020!? Doesn't Mr Fauchi know the meaning of the word 'surprise'? Or, should we all be amazed and impressed at his prophetic, precognitive skills? Maybe we should all be asking ourselves how and why no-one in authority appears to be stitching this CDC/NIAID/WHO/UN/GPMB/ Bill & Melinda Gates Foundation & World Bank narrative together, and coming up with the glaringly obvious conclusion that all of this has been pre-planned?

Still not convinced? Well, not only have there been a number of similar 'Event 201' symposiums that predict and plan for a global pandemic, but a 2010 publication by the Rockefeller Foundation contains yet another prophetic article entitled: *"LOCK STEP: A world of tighter top-down government control and more authoritarian leadership, with limited innovation and growing citizen pushback."* One section is worth quoting (as shown in the Plandemic video) where another visionary prediction reads:

"China's government was not the only one that took extreme measures to protect its citizens from risk and exposure. During the pandemic, national leaders around the world flexed their authority and imposed airtight rules and restrictions, from the mandatory wearing of face masks to body-temperature checks at the entries to communal spaces like train stations and supermarkets. Even after the pandemic faded, this more authoritarian control and oversight of citizens and their activities stuck and even intensified. In order to protect themselves from the spread of increasingly global problems—from pandemics and

164

transnational terrorism to environmental crises and rising poverty—leaders around the world took a firmer grip on power."
[xli]

And this is only a snippet from the 2010 Lock Step Report which goes on to predict with uncanny accuracy practically all of the elements of the Covid-19 phenomenon including many of the themes espoused by Dr Klaus Schwab and Co., and the Davos clique at the World Economic Forum whose 'Fourth Industrial Revolution' and 'Great Reset' narratives align at least theoretically (and quite precisely) with the New World Order concept. So, was this Lock Step Report just an insightful prediction or was it a plan? Because if it was just an insightful prediction just like Clade X and Event 201 were reported to be, and if it was genuinely inspired by altruistic, humane motives, then why haven't our nobly-motivated leaders prepared for the worst – in the public interest – based on the conclusions? After all, after putting so much time, effort and money into dreaming up these uncannily-predictive scenarios, one would have expected them to take at least *some* notice of their collective findings – right? Especially as all of them align so neatly together. Otherwise, what was the point of writing the Rockefeller-sponsored Lock Step Report; of the John Hopkins University-sponsored Clade X seminar; and of the Gates-sponsored Event 201 if no-one was going to take any notice of the terrifying predictions?

But on the other hand, if the plan is totalitarian control of the population and these preparatory scenarios were precisely for that purpose, well, it seems we may owe the Cabal an apology, because it seems that they did indeed prepare for what's happening (although they forgot to explain that to us of course) and therefore, things are in fact going more-or-less exactly according to plan.

The Fauchi Dossier then exposes the UN-based World Health Organisation as being in a unique global position to oversee the health of humanity. Unfortunately it is funded in the main by massive contributions from Big Pharma, certain named foundations and national governments. Now-familiar names trip off the tongue such as; Merck; GAVI; Novartis; Sanofi; Glaxo-Smith-Kline; Pfizer; Bayer; Johnson & Johnson; Moderna; Eisai; & The Bill & Melinda Gates Foundation (who else)!? The point is clearly made – just as we explained in Part One – that people are being appointed to senior corporate positions and high socio-political office not on personal or professional merit, but very worryingly, because of their willingness and ability to do what the Big Boys want – whatever the cost. Tedros Adhanom Ghebreyesus' disturbing past before he was appointed the Head of the WHO in 2017 is highlighted; begging the question as to whether he was chosen as the new Director General *because* of his terrorist background and his political role in (allegedly) engineering and then covering up cholera epidemics in Ethiopia, or, that he was chosen *in spite* of this? Maybe being openly backed by Bill Gates, the Clinton Foundation AND by China explains the unprecedented number of votes cast by our representatives at the UN for a new leader of the world's leading health authority who is not only NOT a medical doctor or scientist, but who has also been formally accused of genocide by his own Ethiopian Government? How and why an appointment of this importance can be made to such a key agency of the United Nations without some transparency as to how all these issues were somehow overlooked by the UN is clearly a vindication of the premise laid out in Part One that the Cabal will look after its own.

Returning to the architect of the whole, 'let's vaccinate the globe ideology' we note that Bill Gates' father was a politically-connected lawyer and lifelong eugenicist who served on the Board of Planned Parenthood, so it is not unreasonable to conclude that these 'population control' and 'population enhancement' ideas – along with an intimate awareness of how 'the system' works – would have been with Bill Gates since childhood. In isolation, having any set of beliefs or a particular philosophy on life is one's own private concern of course. But it is when those ideas are combined with big, big money that we start to understand how and why Bill Gates is treated as if he is a Head of State or some such; not because his ideas are necessarily all that great – or practical – or indeed moral – but simply because he has lots and lots of money, and is willing to spread it around. When most of our leaders are driven by the quest for profit, and when we see that most of them have ascended to their lofty positions because they are adept at the Great Lie, then it's a simple mathematical 2 + 2 process to see that any ideology, plus enough money, plus greedy, narcissistic people in positions of power and authority will inevitably result in that ideology becoming real. And Bill Gates has been pushing the, 'let's vaccinate the world' ideology for a long, long time now.

At a Technology Entertainment & Design Conference in 2010 (TED) Gates remarked: *"The world today has 6.8 billion people. That's headed up to about nine billion. Now, if we do a really great job on new vaccines, health care and reproductive health services, we could lower that by, perhaps, 10 or 15 percent."* Whilst we must of course acknowledge the possibility that Gates is a genuinely-motivated philanthropist who is driven by a moral agenda, we must also register the fact that he is a very clever person indeed, with abundant experience of moving in elite business and political circles and *must* therefore know the calibre and motivations of most of those he is rubbing shoulders with. He is also very familiar with the architecture of the establishment and how to grease its wheels, and one cannot be involved in these circles for too long before one realises and understands – and is then drawn into – the Great Big Lie. In other words, billionaire or not, or altruist or not, Mr Gates must be fully aware by now that he is funding a network of sociopaths and psychopaths with compromised morals and severely-handicapped consciences, who, arguably, would be far more suitable candidates for his life-shortening injectibles than the pitiful peasantry whose lives have been evaluated and assessed by Mr Gates and his collaborators only as to their intrinsic monetary 'worth' in a dystopian future. The fact that they all knew that this crisis was coming well in advance but did nothing to genuinely help or prepare the people unless it resulted in profits for themselves – such as awarding uncontested contracts at the public's expense for Personal Protective Equipment (PPE) to cronies of the establishment

– proves that we are NOT dealing with moral altruists or genuine empaths, but as we keep on repeating, with soul-less sociopaths and psychopaths who will prey on the vulnerable and the confused without any remorse or regret as long as we keep inviting them to do so. We simply must be alert to the lie, because that is one of the first indicators and warnings. In April 2020 for example when Bill Gates said on camera that, *"We didn't simulate this, we didn't practice.."* he was lying. When he continued by saying, *"..so both the health policies and the economic policies; we find ourselves in unchartered territory"* – he was lying again. He also lied and obfuscated throughout the USA vs Microsoft anti-trust trials during the late 1990's. So, whatever his claims to altruism and good intentions, the one thing we *do* know is that Bill Gates is a very capable and convincing liar. If we consider the value of all of his billions against true humane values then we begin to see how society's perception of 'value' and 'worth' has become utterly corrupted and how Mr Gates and his oh-so-wealthy co-conspirators would actually be amongst the most impoverished of souls in any genuine humane society – or in any true Kingdom of Heaven on Earth.

The fact alone that Bill Gates claims to be a philanthropist whilst simultaneously acknowledging that his investments in 'public health' have more than doubled his fortune, with his investments in vaccine companies returning 20-fold profits to the B & M Gates Foundation leaves us asking how and why everyone with a shrewd eye on the almighty dollar isn't jumping on the philanthropy bandwagon so that we can all save the world whilst at the same time making billions in personal profits? Again, something doesn't ring true here. Genuine altruism has the effect of diminishing one's personal fortune – not increasing it. The next thing we'll hear is that Big Pharma are producing all of these artificial medicines, drugs and injectibles out of the goodness of their blessed little hearts, and that the profits being generated are just one of the unexpected benefits of their noble altruism and heartfelt concerns for the genuine health and wellbeing of the people. Yeah. Right.

The more we hear about the breadth and depth of the Gates Foundation empire, the more we should be concerned. With its tentacles in so many diverse-yet-interconnected pies, including 70 of the worst-polluting businesses in America and with other commercial organisations who deal in child-labour and other discredited and amoral practices that are hypocritically denounced in public in order to maintain the façade of politically-correct propriety, it seems that no moral or ethical breaches are off the table if they serve to advance the agenda of the Cabal. I will leave the reader to research how and why Mr gates is buying up vast tracks of arable land in the Unites States; or why the Indian Government has banned the Gates Foundation from operating there after it was discovered that many thousands of tribal girls (mostly uneducated, impoverished and unaware) were in effect, being experimented upon in 2009 with disastrous results for many of them.[xlii] The additional 500,000 children who became paralysed between 2000 and 2017 after another Gates-sponsored experimental polio vaccine hardly bear mentioning, because after all, these people are too poor and too disempowered to be able to sue for compensation. No moolah? No problem! Experimental research on uneducated, gullible children without their full consent? Now where have we heard that type of thing happening before? Wasn't that sort of thing outlawed and forever prohibited after the Nuremberg Trials?

The anti-trust laws of the USA are very explicit – as are similar laws in all developed countries – as to 'insider trading' and business monopolies and conflicts of interest that are either unfair or that harm the public interest, yet the Dossier goes on to explain that over 1,300 patents are being used by this Gates-sponsored network of interconnected agencies and institutions – many of which are 'parked' in seemingly non-affiliated businesses – to profit illegally in violation of the anti-trust laws. Why no-one in authority seems to care enough to want to do something about this is a question that is left hanging – and one reason why we in Ireland set up our own *Peoples Tribunal of Ireland* in July 2020 in face of the repeated and persistent failures and refusals of 'the authorities' – including the Irish Courts – to raise a finger against any of their colleagues in the Cabal.[xliii] It seems a worldwide phenomenon does it not, that 'the Rule of Law' which is the concept upon which all of our democracies are supposedly built, does not in fact apply to the Cabal. This would of course make a nonsense of all of our respective Constitutions that are rooted in the proper application of the law, according to the principles of the Rule of Law. In other words, our various justice systems, our police authorities and our domestic Courts are only willing to apply the law up to a certain point. The peasantry (that's us by the way) can expect to have the proverbial book thrown at us even for minor infractions, but the corrupted amoral elites who actually personify the definition of 'a criminal organisation' even in their own words? Well, we know that it may not align with the Constitution and that technically, yes, these people are committing crimes essentially as a career choice; but the peasantry simply has to understand the realities here and how awkward it would be at the Golf Club or at public events or at private dinner-parties for the Cabal NOT to be able to continue the pretence that they are somehow superior to the rest of us. Criminal convictions and such like do not roll off the tongue easily in pretentious circles, and do not sit well in a public narrative that urges the public to 'trust the authorities'. What we actually have is a bought-and-paid for (unacknowledged) exemption for the Cabal and their cronies, which means that we do NOT have law *per se* – certainly not in any logical, rational or indeed legalistic understanding of the term. What we have here is criminal protectionism under the guise of government, and it's a lie that each of us becomes complicit in whenever we knowingly engage with them. So, why then, if we wish to consider ourselves persons of truth and integrity, do we continue letting these criminals run the show?

The fact that Big Pharma has been foisting scam after ruse on the unwitting public for decades, and have been selling placebos, dangerous drugs and injectibles in full knowledge that they could cause injury and death was no more than a statistic to be taken into

170

consideration as they weighed up the profits to be made vs the fines to be paid, as-and-when they were caught out – or, as-and-when some authority figure with the gumption to so – tried to do something about it. When their accountants informed Big Pharma that the profits to be made by creating and selling dangerous products outweighed the cost of punitive fines – well, then there was really no more to discuss was there, other than how to 'manage' the fallout so as to ensure the bottom line wasn't too badly affected.

This is how evil manifests itself in our society today in the cosy cartels of professionally-placed criminals and their protectors in the Courts; and via the warm and soothing tones of polished adverts reassuring us subliminally that we can of course trust the Big Boys and Big Pharma in particular because after all, they are the born leaders and experts who know what's best for us. Again, it all comes back to money and the realisation that if you have enough of it you can buy almost anything – even the moral and ethical standards of the medical profession; of prestigious schools and universities; of scientific establishments and of political collectives with their stooges and cronies who have clearly lost any true concept of 'value' in humane terms. The wholesale corruption of the scientific-medical sphere by the big players was captured in a recent documentary that exposed all of the dodgy practices being engaged in by Big Pharma and being actively facilitated by far too many compromised medical and legal professionals as they conspired together to achieve the ultimate goal; that of securing ever-more profits by any underhanded means – chiefly, as always, through the active deception of the public.

The Rockefeller Connection. Perhaps the greatest wholesale medical deception of our time was funded by John D Rockefeller in the early 20th Century when he married up the burgeoning petroleum industry with so-called 'modern medicine' (comprising petroleum-based drugs and vaccines) and, by bestowing conditional grants to Colleges and Universities then shifted public consciousness and opinion so as to perceive age-old natural remedies and cures as something that was now quaint and outdated. Rockefeller had seen a great business opportunity if only he could align his existing business interests with 'public health' and thereby create a huge, patent-generating profit-making conglomerate that would eternally fund itself. The key was to be able to create new medicines from petrochemicals which, because these were new emerging technologies at the time and were not 'natural' products, they could in fact be legally patented. Natural things such as animals or plants or microscopic elements of nature can NOT be patented precisely because they are 'of nature' and belong to all of us. This is an accepted international law and important principle which has a direct bearing on the Covid-19 phenomenon to be covered shortly, and one of the reasons why the debate about Genetically Modified Organisms (GMO's) has been hot and heavy with many people questioning the ethics and propriety of companies like Monsanto 'creating' and patenting genetically-altered plants and seeds for example, so as to be able to dominate world food markets. Three guesses as to which major international philanthropic Foundation invested a reported $23 billion dollars in Monsanto (now owned by Bayer Pharmaceuticals) in 2010? [xliv]

Rockefeller understood that traditional natural remedies did not tend to be harmful nor addictive, but as we can see, most modern medicines are. In fact, just as MacDonald's uses special additives to ensure a hungry public keeps coming back for more, so does the business model of Big Pharma incorporate all manner of inducements, and coercions and subtle psychological persuasions to ensure that we remain just sick enough – or at least, just scared enough of becoming terribly sick – to ensure that we keep going back for more. And incidental or not, there will be no great rush to remove addictive properties in medicine or even take dangerous drugs off the market as long as there's money to be made.

The Rockefeller Foundation's involvement in the controversial 'experimental vaccination program' of US soldiers in Kansas in 1918 has recently surfaced as a possible cause of the subsequent Spanish Flu pandemic that killed up to 100 million – something that would entirely fit the profile of a megalomaniacal psychopath who saw people as mere units of profit to be experimented upon at will. Whether or not the Spanish Flu was preplanned as a global cull may remain one of those closely-guarded and not-spoken-about topics, at least during these equally suspicious Covid pandemic times.

But Rockefeller wasn't quite done yet. After making sure that any outspoken opponents to his control of the public health debate were silenced, vilified, fired and jailed (sound familiar?) he then suppressed the evidence that many of these newly-patented drugs were actually causing cancer, by shrewdly setting up the American Cancer Society so as to be able to control *that* particular narrative as well. See how it's done? With enough money and an absence of scruples, anything is possible it seems. If there is such a thing as 'a diabolical monopoly' then John D Rockefeller can rightfully claim to be its inventor.[xlv]

Bill gates it seems, is going for Rockefeller's record, with his involvements in the SCoPEx Project, surveillance satellites, Covid-19 passports, micro-needle patches, quantum-dot tattoos, biological cryptocurrency sensors, and genetically-modified serum-delivering mosquitoes – to name but a few. (Yes, that's real too I'm afraid). The SCoPEx Project (Stratospheric Controlled Perturbation Experiment) for example, is that tin-hat conspiracy that we all keep hearing about that is constantly being debunked by the fact-checkers, where the stratosphere is strewn with sunlight-blocking particles in order to affect the weather and prevent global warming. An interesting possible solution to ecological issues perhaps, but should we be really trusting Mr Gates with it just because he is richer than sin?

Bill Gates' personal relationship with Geoffrey Epstein and with other big names from business, media, politics and from amongst the privileged elites is not something we need to go into here other than in context of the reality that evil partners evil in many unreported and unacknowledged forms. [xlvi]

CHAPTER FIFTEEN

THE FACTS DON'T LIE

FACT OR FICTION – WHAT DO WE KNOW, NOW?

In an effort to avoid overly-complicated arguments and explanations about why we should be raising questions about the whole Covid-19 phenomenon at all, and whether or not we should be increasingly suspicious of the official narrative, here are some truths and realities which should speak for themselves. Please remember the fundamentals of 'The Great Lie' as discussed in Part One, especially the crucial differences between empaths and sociopaths, and the fact that whilst individuals possess a conscience that drives moral decision making, that profit-driven corporations, government agencies and institutions, and other similar large organisations do not.

THE FACT-CHECKERS. Many of the self-declared 'truly independent' Fact-Checker Agencies are funded by Big Tech, by corporate interests, by national governments, by political parties, or by individuals who are actively driving the Covid-19 agenda itself. Staff in those agencies have often typically held positions with 'the establishment' including with the respective governments or mainstream media, or, who have longstanding lucrative contracts that bind them to the official narrative. Others, such as the now publicly-discredited Dr Peter Daszak, President of the EcoHealth Alliance non-profit organisation who, while being funded by the Bill & Melinda Gates Foundation (BMGF) was (improperly if not illegally) channelling moneys for dangerous coronavirus research to the Wuhan Institute of Virology – yes, the very same place where the virus originated; that Dr Daszak was then hired by the WHO (also underwritten by Bill Gates) as the US lead for a team of experts sent to investigate the possibility of a lab-leak in Wuhan, and was then actually co-opted onto Facebook's official fact-checker team despite these glaring conflicts of interest. And it's not as if the WHO and

Facebook weren't aware of Dr Daszak's connections to the Wuhan Institute of Virology and the Gates Foundation, because any simple search online connects the EcoHealth Alliance with Gates & Co., as well as with many other key players such as Dr Anthony Fauchi, medical advisor to the U.S. President and head of the National Institute of Allergy and Infectious Diseases ('NIAID'). A simple Google search unearths a 2016 video report where Peter Daszak boasts about the deliberate manipulation of coronavirus S1 spike proteins at that Wuhan Institute by "his colleagues" where he notes specifically how any such "nasty pathogens" created in a lab could be potentially catastrophic to mankind.[xlvii] So how on earth did Peter get selected as a fact-check partner for Facebook posts that referred directly and specifically to his own clandestine work, whereby he and his toadies could censor and bury anything that pointed at the troubling emerging truth?

Professor Margot Susca would be another example inasmuch as she was an ardent Clinton supporter and vehement anti-Trumper whilst being paid by Facebook to qualify other fact checking agencies, who in turn would then secure lucrative FB contracts. As a result, we had fact checkers with a clear political agenda controlling the world's largest social media platform during a U.S. presidential election. And there are many, many more examples of fact-checking agencies springing up in clusters to serve the Big Boys, not in response to any real quest for truth, but simply, again, because of the moolah – the lucrative contracts. Remembering Chapter Seven and our discussion regarding the particular skill-sets required to be able to mask the true objective of one's work by pretending that one is doing something different altogether? This, generally speaking, is the model at play here. The real objective is to further the official narrative whilst suppressing, burying or ridiculing contesting claims – even if they are true. But the public is being told that facts are simply being checked and corrected, so that the public isn't misled. Lies upon lies once again, and all done for short-term profit. It therefore appears that 'fact checking' and 'debunking' of inconvenient information can be a very lucrative business these days, and it doesn't really seem to matter how honest-or-dishonest, how objective-or-subjective or biased-and-opinionated, or prejudiced, or politically-motivated or professionally compromised the fact-checker may be, as long as they know the script – and they stick to the script.[xlviii] The whole notion of fact-checking as some sort of protection from 'fake news' that incorporates suppression and censorship is just another covert attempt by the establishment to displace our own intuitive ability to each discern truth from falsehood, by once again, institutionalising and formalising another crucial aspect of being truly human; that of the ability to think for ourselves and decide upon the evidence and facts before us, as to where the real truth lies. Hence, fact-checking by others 'on our behalf' is an affront to our intuition and to our intelligence, and is yet another example of the intentions of the establishment – in yet another of its myriad interconnected forms – to totally dominate our minds and thoughts, as well as our bodies and spirits.

Some of the more prominent Fact Checkers who are clearly pushing the official narrative and suppressing and burying contrary or 'inconvenient' material at the expense of objective truth and of quality investigative journalism include (but are not limited to) these individuals, agencies and information providers, whom we name and expose here for the simple purpose of raising a very large question mark over the credibility or authenticity of anything they print. This is on the basis of two factors: (i) That they are directly associated with, or are linked to and/or are being paid by major players in the Covid-19 official narrative; and/or (ii) that they have demonstrated repeat bias and a lack of objectivity in their 'fact-checking' duties. For example, the Mayo Clinic online projects an image of medical professionalism opening its Covid-related myth-debunking commentary with the unambiguous statement that, "Currently, no cure is available for Covid-19".[xlix] Then, after trashing any-and-all other possible claimed reliefs, it directs the reader to, "Go get the vaccine." But why, Mayo Clinic? Didn't you just admit that there's no cure? In other words, keep this list handy as you read the latest 'news' or these agencies' fact-checking rebuttals of whatever contrary claim they might be trying to invalidate, and please weigh up *their* overt and covert conflicts of interest, and *their* persistent alignments with 'the MSM agenda' and with the establishment as you discern the quality, purpose and intent of their so-called 'research'. The list is only a very small sample of those outlets where we personally have sufficient proofs to question their objectivity and integrity, so please understand that this is by no means an exhaustive or comprehensive list. With respect to the possibility that some people who work in these agencies are themselves being lied to, manipulated and misled, here is a short sample list of already-discredited fact-checking and 'debunking' agencies, several of which are funded by the B & M Gates Foundation, by Big Tech, by Big Pharma, by national governments or by the compromised MSM.

News & Media Outlets
Reuters
The Journal.ie
Facebook fact-checkers
Poynter Institute
Wikipedia
USA Today
The Irish Times
RTE News & Prime Time
Euronews
BBC News
The Western People
The Associated Press
Russia Today
France 24
Roscommon Herald
Mayo Clinic

Specialised Agencies
Science Feedback
Fullfact.org
Healthfeedback.org
Snopes.com

Politifact
Bosphorus Global
Fact Checking Turkey
Factcheck.org
AfricaCheck
AP FactCheck
Sci facts
Nature Medicine
Fact Check Armenia
News Guard
Mediekollen
Check Your Fact
Decrypteurs
Lead Stories
The Conversation
Media Wise
The Dispatch
(Please add...)

If that shortlist (with the emphasis on it being far too short) is still too long to bother investigating, then just peruse the UK's Fullfact.org 'meet the Team' webpage as just one example of a highly-professional setup that is devoted entirely to fact-checking, and see how many Lords and Ladies or ex-government, media or big business personnel you can spot amongst the 40-or-so bios on display? No insinuations are being made here other than establishing the undeniable connections with the establishment and the likely bias that would travel with that. Perhaps if the Fullfact.org Team were to tackle some of our *Integrity Ireland* publications with a view to confirming the truths therein, we might be a touch more convinced.[1]

There is no doubt that many fact-checking teams are engaged in extensive research and that some of that work is valuable in eliminating contrived or crackpot theories. But the real problem lies in the unrebutted claim (which has NOT yet been debunked) that their sole reason for existence is to protect the official narrative – for which work they are getting very handsomely paid – right? There is no question that most of them are very articulate and exhaustive in their reports, and that the overall sense one gets when reading their rebuttals is that of rational and sensible professionals coming to the logical and reasonable conclusion that anyone who doesn't align with the official narrative is obviously a crank, a fraud or a nutty conspiracy theorist. Sometimes they are right of course. But more often they are not. It takes a forensic understanding of what they are doing and how they are doing it – as well as more time than most of us can spare – to wade through all of the referenced materials to uncover what is, and what is not, ultimately 'true'. Because one of the fact-checkers' core tactics, just like many politicians and official Covid-19 spokespersons, is to simply blather on convincingly about the purported facts-and-data that have been confirmed by this-or-that trusted source, throwing just enough truth into the mix to give the impression of verified fact, and of properly-conducted, objective and independent research sufficient for us to conclude that of course, it must be true. But maybe it's now time to start fact-checking the fact-checkers and debunking the debunkers before we are all driven completely insane?

As discussed in Chapter Two, the sad and unfortunate reality is that with very few and isolated exceptions, the mainstream media (MSM) has long since been wedded to the establishment and the information we are getting from them cannot even loosely be referred to as 'news' any more, not when other terms such as propaganda, social conditioning and lies are far more apt. It's a truly depressing realisation; that most of the institutions and organisations that dominate human society are simply NOT grounded in moral principles nor even upon ethical standards, but rather (at very best) in the mere *pretence* of the same, for the purposes of advancing contrived agendas that are deceptive, exploitative and dishonest at their core. It is nevertheless crucial that we accept this reality if we are to tackle the Covid-19 phenomenon in any real and effective way that maybe, just maybe, could result in a general awakening of the people as to how we might use this awful reality as the basis for a better world.

Conspiracy Theorists. Possibly one of the most enlightening and surprising elements of the Covid-19 phenomenon (for me personally at least) has been the conversion, in the awareness of the general public, of so many conspiracy *theories* - into *proven* conspiracies. And this of course, is why the fact-checkers and debunkers are so very hard at work. This is also why those of us who have been speaking up and speaking out against the evils of the establishment (or at least the wicked aspects of the Cabal) have also become targets for vilification, character assassination and derogatory reports in the mainstream media, because the establishment is desperate to prevent that truth getting out; that there is indeed a massive ongoing conspiracy by interconnected amoral elites to deceive, defraud and disempower the people for the Cabal's own ill-gotten gains. Now this is not to say that we do not have a sizeable number of (possibly) well-intentioned but poorly-informed people out there who are adding 2 + 2 together and coming up with wild and implausible speculations that have not been properly thought-out or empirically researched; but this is an understandable human reaction to the need to understand things that seem to be beyond explanation unless we can identify a malicious conspiracy by Freemasons, the Illuminati, the Jews, or by alien reptiles masquerading as world leaders for example. My own explanation of course is that we do not need to be inventing outlandish theories or clandestine gatherings for that which is staring us right in the face. Because no matter what form or shape these collectives might take, the conspiracy *per se* is actually one of good vs evil and of truth vs lies, and we need to properly understand this and be able to clearly differentiate between the two. This also goes for our treatment of 'conspiracy theorists' or of those (such as myself or similar others) who present 'alternative' explanations and reasons for the Covid-19 phenomenon, noting how hypocritical it would be NOT to insist that the same exacting standards that we are applying in this research should also be applied to this book, or indeed to any-and-all sources of Covid-related information. For this reason we should be especially alert that in our enthusiasm to 'prove' a certain malicious conspiracy by this, that-or-the-other group, that we do not drift from an objective appraisal of those who posit alternative views asking for instance; what are their personal or professional motives? i.e. what have they to gain from pushing any particular theory? Because if they are making money, or are advancing political ambitions or even just making a big name for themselves as opposed to having a genuine, proven mission to share the truth, then we should be alert to these influences. An individual's track record is usually a good indicator of their motives and intent. This does not necessarily mean that such people are NOT telling the whole truth – just that we need to be discerning and objective about who, what, why and how we believe.

Testing for viruses. Before any scientific test is considered reliable, a 'gold standard' (not the monetary one) or a *'Criterion Standard'* is established as a benchmark to measure the efficacy and accuracy of that test.[ii] With new or evolving diseases (such as Covid-19) that 'gold standard' cannot be properly established until sufficient knowledge has been gathered by the scientific community whereupon a precise test is then created for that specific purpose. In the absence of any such specific SARS-Cov-2 test, and on the public urgings of the World Health Organisation in March 2020 to all countries to, *"test, test, test"* the medical community adopted the pre-existing RT-PCR test as its (temporary) gold standard.

The RT-PCR Test.

Arguably, anyone who properly understands the place that the PCR Tests have played in the Covid-19 phenomenon must, unless they are mentally challenged; are irredeemably obdurate; and/or have been system-conditioned to the point of brainlessness, come to the conclusion that the whole sorry Covid event from the beginning has been one massive, premeditated and calculated fraud on the people – and a fraud with far-reaching and very sinister consequences.

But first, a few necessary facts. The 'real-time reverse transcription polymerase chain reaction (rRT-PCR) test' was invented by Nobel prize-winning scientist Dr Kary Mullis who, somewhat conveniently for the Cabal, passed away just a month before the Covid-19 alarms were set off. This didn't prevent him from getting a few digs in at Anthony Fauchi before he passed on, implying that Dr Fauchi was no more than a political mouthpiece who lacked the clinical knowledge to understand the PCR test – or, that he was deliberately fogging it!?

Anyway, the PCR process 'works' by basically multiplying and magnifying microscopic materials until sufficient 'evidence' is revealed, but it is important to note that even Dr Mullis himself said that the PCR *process* was NOT a diagnostic test – something since repeatedly (but not yet convincingly) 'debunked' of course, by many of the now-discredited fact-checkers from the previous list.

A simple way to understand how the PCR Test works is to imagine using a Google online map to identify fields where there is a plague of mice. In this sense, 'a plague' means say, 10 mice per square metre. But obviously, you can't see any mice at all until you zoom in the camera say, 25 times – sufficient for the mice to become visible enough to count. The mice in this case represent the virus. The basic problem with the PCR Test was that first of all, if the tester saw even one mouse in the field it was being reported as a 'positive' result and thereby implying 'an infection' (or plague). What's worse though, is that other testers were zooming in over 40 times, magnifying the field to such an extent that one could find all sorts of 'evidence' of mouse-related activity such as burrows or droppings, and this too was getting registered as 'a positive result' – or as evidence of a plague-like infestation. Obviously this is a very simplistic explanation, but should help those of us who are scientifically-challenged to get the picture.

The apparatus used to conduct these Covid-related PCR Tests also has differing ranges of magnification, meaning that a machine in one location may pick up 'evidence' of a virus that is not detected by another. Furthermore, the manufacturers of the PCR test kit warn explicitly in the accompanying literature that at least 4 other existing viruses (other than SARS-CoV-2) could trigger a 'positive' result. These are HKU1; NL63; OC43 & 229E, otherwise known as common colds that cause mild to moderate upper-respiratory tract illnesses, and, if the host has any of these viruses – or was ever *previously* exposed to them apparently (think mouse-droppings in a large field) – then the test could then register a 'false-positive' for Covid-19 – which is then reported as another Covid 'case' – right? [lii] The higher the number of test-cycles (zooming-ins) the higher the % chance of false-positives being returned. In other words, those in the know including the World Health Organisation, the CDC, the NIH, the NHS, the HSE, Dr Anthony Fauchi and Bill Gates for example, all knew that the so-called 'results' from these PCR processes (not real tests at all) were not only unfit for the task, but were producing totally false and misleading results.

This false-positive anomaly relates in turn directly to another important scientific factor known as the 'margin of error' – a statistic that informs the tester of the accuracy of the outcomes in any given medical/scientific test in respect of the numbers tested. A typical acceptable margin of error in scientific tests would be around 5%, meaning that if 100 people were tested for example and 4 of them registered as 'positive', that because the margin of error in that particular test was set at 5%, that it is possible, likely and even probable that all of those 4 'positive' results were in fact 'false-positives' or errors of data. Any such test that strays above a 10% margin of error is in effect, not really a credible, analytical test at all but more a quasi-scientific procedure that produces unreliable and relatively meaningless, and purely speculative data. In respect of Covid-19, the PCR Test had already proven notorious for returning an unacceptably high number of 'false positives' (up to 33% in some repeat systematic tests) in circumstances where different countries, agencies and organisations *still* did not have a reliable standardised gold model for testing for Covid-19.[liii] In short, the PCR Test is wholly inadequate to the task and falls way below the requirements of scientific testing standards, begging the obvious question as to why the WHO, the CDC, the NIH and Co., insisted that *only* the PCR procedure was to be used for Covid testing?[liv]

Ethylene Oxide: Another more recent and very disturbing discovery is the fact that self-administered Covid-19 Rapid Antigen PCR 'EO' Test swabs are, for some inexplicable reason, coated with Ethylene Oxide which is described on the US Government's NIH Cancer website:

> *"At room temperature, ethylene oxide is a flammable colorless gas with a sweet odor. It is used primarily to produce other chemicals, including antifreeze. In smaller amounts, ethylene oxide is used as a pesticide and a sterilizing agent. The ability of ethylene oxide to damage DNA makes it an effective sterilizing agent but also accounts for its cancer-causing activity.[lv]*

Wikipedia hosts a much more detailed article, from which we have pulled the following short quotes: (American spellings).

Ethylene Oxide is / affects / causes:

- It is industrially produced.

- It is too dangerous for direct household use.

- It is a very hazardous substance. At room temperature it is a flammable, carcinogenic, mutagenic, irritating, and anaesthetic gas.

- It is so flammable and extremely explosive that it is used as a main component of thermobaric weapons.

- It has irritating, sensitizing and narcotic effects.

- Chronic exposure to ethylene oxide is also mutagenic.

- It is a proven carcinogen.

- It is associated with breast cancer incidence.

- An increased incidence of brain tumors and mononuclear cell leukaemia was found in rats that had inhaled ethylene oxide.

- There is evidence from both human and animal studies that inhalation exposure to ethylene oxide can result in a wide range of carcinogenic effects.

- Ethylene oxide is toxic by inhalation.

- It irritates mucous membranes of the nose and throat

- Higher doses cause damage to the trachea and bronchi, progressing into the partial collapse of the lungs.

- High concentrations can cause pulmonary edema and damage the cardiovascular system.

- The gas is already at toxic concentrations when it can be smelled.

- In view of these insidious properties, continuous electrochemical monitoring is standard practice, and it is forbidden to use ethylene oxide to fumigate building interiors in the EU and some other jurisdictions.

- Ethylene oxide causes acute poisoning, accompanied by a variety of symptoms. Central nervous system effects are frequently associated with human exposure to ethylene oxide in occupational settings.

- Headache, nausea, and vomiting have been reported. Peripheral neuropathy, impaired hand-eye coordination and memory loss have been reported in more recent case studies of chronically-exposed workers at estimated average exposure levels as low as 3 ppm.

- Ethylene oxide penetrates clothing and footwear, causes skin irritation and dermatitis with the formation of blisters, fever and leukocytosis.

So... why is this stuff being put on a cotton swab that is going right up people's noses as far as the hematoencephalic blood-brain barrier (BBB) which is already know to be susceptible to 'the effects' of Covid-19?[lvi] Despite persistent efforts by fact-checkers to debunk the possibility that the deep nasal swab PCR Test *cannot* damage the blood-brain barrier, an October 2020 article in the *Daily News* reads:

> *"In everybody's worst COVID-test nightmare made real, a woman's brain fluid leaked out of her nose after a coronavirus nasal swab."*[lvii]

More disturbingly, why is this Chinese-made product the standard Covid test that is being used on children? Surely all one needs to complete a nasal swab is something that is absent any other chemical agents so as to be able to simply collect some spit or nasal fluids without putting the subject at mortal risk - right? Some clean cotton buds or wooden spatulas that can be sealed in vials or plastic bags perhaps? Would it be too much to raise suspicions that these so-called PCR 'EO' Tests are another part of the diabolical plan?[lviii] Could this possibly be the reason why the WHO instructed the authorities NOT to look for the specific SARS-CoV-2 virus at all, but to rely instead on the RT-PCR test – a 'test' which we now know is generic to any coronavirus debris present-or-past and, as far as collecting accurate diagnostic data is about as useful as an ashtray on a motorbike?[lix]

The Truth About Face Masks

It has been a long-accepted social practice is Asia for people to wear a face mask if they know they have a cold or flu, and must nevertheless go out in public. This was an admirable act of voluntary social responsibility so as to prevent the wearer from accidentally infecting others when they coughed or sneezed. It also sent a visual warning that the wearer was sick, and that others would therefore be wise to keep their distance. The fact that surgeons and nurses also wear face masks, mainly to protect themselves and vulnerable patients from oral and nasal germs, but that surgeons for example, need to replace their masks every 90 minutes or so, brings the whole idea of ordering the population to wear facemasks 'for protection' into some kind of focus. But given the possibility that ordering the population to wear face masks is more an exercise in authoritarian control than it is to do with genuine health measures, there are a few common-sense, straightforward questions that we should all be asking, including: Why for example, are we wearing face masks for *this* particular coronavirus, when we have never done so before – at least not in the Western world – especially in light of the fact that we now know for sure even by their own official data, that Covid-19 (or whatever it really is) it is NOT any more dangerous than the seasonal flu? Furthermore, if face masks are genuinely necessary for our protection, then why aren't the authorities checking the quality and integrity of people's masks instead of just checking whether they are visually complying with the 'face coverings' diktat? After all, there are various types of face coverings out there that provide little or no protection at all, and believe it or not, we have actually seen people putting on masks before a checkpoint, and taking them off afterwards simply to avoid a fine or a reprimand. But, just like not wearing a seatbelt is dangerous – so is not wearing a face mask (according to the authorities) except that risking a traffic accident and being exposed to a virus are two entirely different risks that are utterly incomparable as far as the method for protection is concerned. One deals with a possible incident (a car crash) that happens at a set time and place, and wearing a seatbelt provides protection from injury *only* at that precise moment. But because we don't know if-or-when that moment might arrive, we keep our seatbelts on at all times in a moving vehicle – right? On the other hand, the only way to protect ourselves and others against a virus (we are told) that is supposedly, possibly, everywhere all around us at all times, would be to wear facemasks everywhere, and permanently, from dusk to dawn. Maybe we should also wear masks while we sleep, because having such a deadly virus on the loose that could infect us at any time or place of its choosing is akin to being in a moving vehicle at the imminent point of a potential fatal accident without having a seatbelt on, is it not?

So, either it should be masks on *all* of the time, everywhere, or the whole 'masks-as-protection' concept is a nonsense, as is articulated here in an email from Dr Anthony Fauchi himself, in February 2020:

> *"Masks are really for infected people to prevent them from spreading infection to people who are not infected rather than protecting uninfected people from acquiring infection. The typical mask you buy in the drug store is not really effective in keeping out virus, which is small enough to pass through material. It might, however, provide some slight benefit in keeping out gross droplets if someone coughs or sneezes on you."* [ix]

Need we say more? Furthermore, unless a face mask is vacuum-sealed around the nose, cheeks and chin (similar to a gas mask) then common-sense tells us that something that is up to 100 times smaller than a speck of dust or a red blood cell, yet is intelligent enough to know the difference between the local supermarket where it CAN infect people, and a political function at a golf club where apparently, it can NOT – well, obviously, a virus that intelligent and that discerning is not going to have any problems getting past any loosely-applied mask, scarf, pulled-up roll-neck or plastic screen – right? It seems to me that we should be looking for some other reason as to why 'they' (including Dr Fauchi) are pushing face masks so hard, other than purported 'health experts' being complete and utter morons.

For the non-Irish readers out there, we are referring here to what has become known in Ireland as 'Golfgate' where, the day after legislating social distancing restrictions for the peasantry that allowed only 5 people to gather indoors and 15 outdoors, "because of the grave threat posed by the virus", that over 80 members of the establishment gathered in a golf resort in Connemara for two days of partying, politicising and playing golf – (at the taxpayers' expense of course) – apparently blissfully unaware of any apparent risk from the purported virus, despite many of them being of advanced years and therefore (as they all surely knew) at direct and imminent mortal risk of infection and death.

In fact, Seamus Woulfe, the Supreme Court Judge who had signed the Covid legislation into law in his prior role as Attorney General was there, along with a number of politicians and senior civil servants, vulture fund managers and other dignitaries, with not one of them wearing a mask – or social distancing – raising the all-too-obvious question of what did *they* know that *we* didn't? Given the democratic principle of the separation of powers, it is understandable of course how all of these otherwise officially-separated individuals would jump at the chance to innocently socialise, but it really does deserve repeating: What did *they* know that *we* did not? Is there a dangerous virus out there or is there not? And if there is, then why aren't our politicians and judges taking the same precautions that they are projecting, via stringent legislation and legal punishments, onto us? Could they really be that stupid – or, is arrogance and hubris a protection from the virus after all? (No need to answer that one).

Masked Dangers. Del Bigtree is a prominent figure in the USA reporting online on his High Wire program about many questionable aspects of the Covid-19 official narrative. One week recently, he focused on face masks, and, with a digital air-quality monitor unit to measure CO^2 demonstrated very convincingly that his 11-year old son was inhaling toxic levels of CO^2 whilst wearing a variety of face coverings – some worse than others.[lxi]

Other than the obvious fact that humans are not designed to breathe through fabrics and that there is undoubtedly a health consequence from doing so over extended periods, the only question that remains is whether or not it is even *more* hazardous to go without a face mask – such as would be the case if we were exposed to smoke or gas, or to some invisible, super-intelligent virus that is not in any event going to be deterred by fabrics whose weave is multiple times larger (in most cases) than the reported size of the virus molecules themselves.

The size of the virus is important. Understanding that there are a thousand microns in a millimetre for example, the width of a human hair is c.50-150 microns, a grain of salt is around 60 microns, a dust particle measures c.10 microns, and the various experts tell us that coronavirus measures around 0.1 to 0.3 microns – which is very, very small indeed. This in turn translates to between 100 and 300 nanometres (nm) with 1,000 nm equalling one micron.

In fact, if we were to magnify a human hair 4,000 times so that it is approximately the same width as a person, then a coronavirus particle would be the same size (at 4,000 X) as a poppy seed. Given that the standard N95 surgical mask has a weave diameter of around 300nm it is clear that single, unattached viruses (at 100-300nm) will obviously be able to pass through even professionally-constructed surgical masks. When suspended in respiratory moisture-droplets that measure between 5-10 microns however, viruses are far more likely to be trapped in a mask that is tightly-woven when a person coughs or sneezes. On the other hand though, unless a mask is vacuum-sealed and impervious, no manmade fabric is going to stop these little critters coming-and-going if they decide to pay you a visit.
lxii

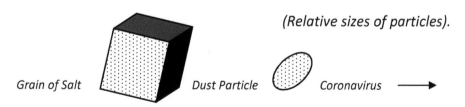

(Relative sizes of particles).

Grain of Salt   *Dust Particle*   *Coronavirus* ⟶

Can't see the coronavirus? Maybe that's the point. At only around 100 billionths of a metre none of us could do so without highly specialised electromagnetic microscopes. But there's another relevant issue to consider as well. When doctors and surgeons wear N95 masks they do so with great care, and change them regularly. So if this is the expert's 'best practice' to prevent them from catching disease, then why aren't we being subjected to the same protocols? Maybe it's because the whole notion is unenforceable, and because the casual wearing of ad-hoc bits of material over our faces is not at all about any genuine attempt to control the physical spread of disease, but is instead all about the need to psychologically control the population?

So, at the end of the day, surgical face masks might offer *some* protection *from* infected persons in clinical close-quarters. They may also help safeguard others if properly worn by an infected person, and perhaps this should be a 'highly recommended' practice for sick people who go out in public, just as they have been doing in Asia for decades. But mandatory face masks of ill-defined construction for everyone on the planet because of some dubious coronavirus that is no threat to most of us and may yet in fact prove to be the theoretical -imaginings-turned-diabolical-constructions of Machiavellian and Frankensteinian actors with a sinister alternative agenda? Well, the choice is obvious for me. Other than to avoid a potential dispute with a business owner or to avoid arrest for non-compliance, or, in circumstances where I know that I do have a cold or flu and wish to do my civic duty, I for one will NOT be wearing one. And I will continue to decline to do so, respectfully your honour, with a warm and cordial, and very visible smile.

SOCIAL DISTANCING
When we come to social distancing as a concept to prevent the transmission of an infectious disease that *cannot* travel, let's say, more than 2 metres, airborne, between individuals, then the idea of social distancing to a distance of 2 metres makes perfect sense. It even makes sense in a scenario where the 2 metre distance prevents

an infected person from getting too close to a healthy person, who then has a 2 metre head-start to flee from the infected person should they start violently coughing and sneezing. But clearly, that's not what's going on here. Because, even putting aside the known fact that whatever Covid-19 is, it is certainly no worse than the common cold or flu, and that most of us would have long-since acquired what is known as 'herd immunity' just as we would with any new strain of the flu; there is the contradictory position being held by the authorities, that depending upon levels of humidity the virus can remain suspended in the air for 3 – 10 hours for example, or that it can live on various surfaces for hours or even days (depending on which 'scientific experts' you may be listening to) which in turn makes a total nonsense of social distancing as a protective measure other than asking those who know that they are infected to please wear a mask and keep their distance. But then again, if we are to believe the official narrative, any or all of us could be morbidly infected with the virus at any given time – symptoms or not – and if you are in any doubt about whether or not you may be asymptomatic, well, you can always trot along to your local Covid-19 Testing Centre – suitably masked and distanced – to avail of the PCR Test which after all, is only 97% inaccurate. But better to be safe than sorry – right? ...Really?

Remembering that practically all of the official information relating to Covid-19 is now being heavily monitored and censored, we sourced research from the 1930's where Harvard Scientists discovered that sneeze particles easily carried across an open classroom that was several yards (metres) long.[lxiii] A more recent video explains that sneeze droplets carry over 27ft (8.2m) in normal indoor conditions and can hang in the air for several minutes even if they are not picked up by the air conditioning, which, as it is wont to do, can and will disperse whatever is in the air around the whole building. So clearly, the 2 metre (6'6") rule is NOT nearly enough distance if the objective is to protect people from near-contact transmission.

Again, little of this 'social distancing' mandate makes sense other than in terms of wanting to disrupt society and unsettle and isolate people as quickly as possible on a global scale. Because if the intention is to gain control of the population through fear, isolation and confusion, then this is exactly where to start. Tell them that a killer virus – too small to see, and only detectible by us experts – is on the loose. Then tell them to trust us, because we are of course working in their interests as we ban all forms of contact-socialising as we test out our control methods on them, to see how compliant they will be. Then, while having a knees-up at the members-only Golf Club we could bring in laws that punished anyone who didn't comply, and the police and civil servants would be reassured that whatever had to happen to the peasantry, that they need not worry because their jobs were safe.

And how will the authorities be able to identify those who are not complying – one might ask? Well, by a curious coincidence, the much talked-about 5G network rollout which was originally scheduled for 2020 deployment and which technology relies on high-speed, satellite-based, wire-free, superfast, cloud-based internet.. will greatly enhance the military's ability for example, to surveil individual enemy soldiers, or terrorists – or dare we say it – non-compliant citizens as long as they are not too closely grouped together. In fact, if we could just get them all to stand about 2 metres or so apart (rounding up to an interesting 6 feet 6 inches) that would be most helpful in helping us to identify, track and monitor any undesirables so that if a decision is then made to send in a drone, or a genetically-modified 'mechanised' mosquito (yes, they do have them too) or just press a button somewhere so as to be able to flood the target with disabling nanodes or mood-responders that were injected during the pandemic? Well, just saying like.. that it would be really, really helpful, that's all, if everyone could stop congregating in dense crowds while we get the 5G system up-and-running properly. Maybe this is the REAL reason why we are being instructed to stay 2 metres (6'6") apart based upon deliberate misinformation about how far disease can supposedly travel in coughs and sneezes?

There are many, many such examples of the establishment talking out of one side of their face about the supposed 'terrible dangers' we are all facing and then acting differently whenever they think the cameras are not on them. The RTE Christmas Party for example, where the broadcasters who have been drilling us all daily – and endlessly – about the need for stricter safety measures, spent a drunken evening dancing, hugging and kissing – and possibly coughing and sneezing together as well – (and who knows what else) along with RTE Management, before realising that someone might be taking private pictures!?

Similarly with our great and noble leaders at the EU, who line up like mini-men on that great Strasbourg stage, all masked and appropriately-distanced for the official photographs before returning to their furtive conversations and Machiavellian exchanges in secretive intimate clutches – absent masks, and close enough to smugly nudge-and-wink – sharing their upper-class germs, safe in the knowledge that the official reporters and photographers very much know their place, and that the public will never get to see the shocking hypocrisy on display. Because then of course, someone might start asking that very uncomfortable question again: "What do *they* know that *we* don't?" Or, "How come all of these senior politicians aren't worried about the virus?" Or, "Why all the pretence and the masquerade of complying with the restrictions that they are forcing upon us, when they themselves are clearly not doing so?"

As we said in Part One, 'career liars and moral deviants' have no issue with deceptions, nor with hypocrisy nor with the great lie. After all, they have each made careers out of it. It takes a certain type of persistent fool to see all of this evidence of wrongdoing by the establishment and NOT come to the obvious conclusion that something is very seriously amiss.

*   *   *

There are two very important facts to be gleaned from the evidence that is staring us right in the face. First of all, there is the farcical notion that a respiratory virus (such as the common cold) that is already 'out there' in the general population could actually be corralled and controlled by politicians by issuing restrictive edicts and social-containment laws which require not only the population, but also the law-abiding microscopic virus, to dutifully comply with.

Secondly, in respect of the 'vaccinations' and the separation of society into the now socially-privileged vaccinated masses, and alienated and socially-proscribed others who have not taken the jab. Why aren't we all asking the simple and all-too-obvious question as to the purpose of *only* allowing vaccinated obedients into pubs and restaurants for instance, when; if they have indeed been 'properly vaccinated' against this mortal threat, then they could not, can not, and are not at any risk whatsoever from an unvaccinated person, whether that unvaccinated person was symptomatic or not - right? So, what's the problem, and why the continued restrictions on the unvaccinated?

One might almost be forgiven for concluding that these injectibles are actually most effective at removing common sense!

CHAPTER SIXTEEN

THE INJECTIBLES

AND THE UNFORESEEN DANGERS OF LITTLE PRICKS

We open on this topic by revisiting the 2016 video made by Dr Peter Daszak who, whatever else he may be – Frankensteinian plotter, schemer, liar, conspirator or not – is nevertheless one of the world's leading experts on coronaviruses. So, perhaps we should pay attention to what he has to say, especially in an unguarded moment? Whilst explaining the potential dangers of bat-related coronaviruses from China for example, Peter Daszak makes the definitive statement at 2m 40secs into the video, that of the various SARS-like corona-viruses that they were experimenting on in the laboratory which had the potential to infect humans, that:

"They are untreatable with therapeutics, and you can't vaccinate against them with a vaccine." [lxiv]

Maybe we need to repeat that? These man-manipulated viruses are 'untreatable with therapeutics' and we cannot vaccinate against them either!? It's hard to know how many more bombshells we can digest without becoming utterly disillusioned and confused as to what expert to listen to and who to believe any more. But I can assure you that those were his precise words. So, either Peter Daszak is completely ignorant of the fact – since ostensibly 'proven' by the likes of Pfizer, Moderna, AstraZeneca and Janssen & Co., who have rolled out billions of vaccines – that we CAN actually vaccinate against SARS-CoV-2 (well, can we or can't we Guys?) or, could it be that something isn't quite right here – yet again?

Let's take a closer look at what Mr Daszak said and see if we can work out what is going on here, because either; (i) somewhat unbelievably, a world-leading expert on viruses wasn't aware of the decades-long research that shows that various 'therapeutics' (such as ivermectin and hydroxychloroquine tablets) have been impressively effective against existing viruses and other pathogens; or (ii) he knew (or at least believed) that any actual physical release of a bastardised virus bio-weapon like this was going to be catastrophic and untreatable; or (iii) he was lying, or at least grossly overstating the risks for reasons we can only guess at, such as Peter getting a bit carried away on the TV with a doomsday narrative wherein he was a main player; or (iv) could he have been simply paving the way for the absolute need for more 'gain of function' research in the face of this terrible viral threat, so that humanity would be ready with an antidote of some sort when the time came, even if that antidote was not a vaccine or a therapeutic solution? Maybe all that really mattered to Dr Daszak was that funding kept rolling in from the U.S. because we now know that he is not at all averse to telling porkies when it suits his interests.

If Peter Daszak *was* telling the truth when he said these type of viruses could NOT be treated or vaccinated against in 2016, then how and why has Big Pharma even embarked on a supposed '90% effective global vaccination program'? Has something substantially changed since 2016 that we weren't told about – or are we not seeing the bigger picture here? Because if Daszak wasn't telling the truth then how could 'the powers that be' leave Daszak in charge of such an important and controversial project while he made unscientific and inaccurate public claims about something he was supposed to be the world's expert in?

Again, something's not right here. When we add in the SARS-CoV-1 vaccine research (c.2003) undertaken by his colleague Dr Ralph Baric of UNC which *did* in fact result in achieving protective immunity *from* SARS-CoV-1 for the animals in the test – but also, unfortunately, caused autoimmunity, which is when the body's own immune system attacks the host's cells – eventually causing death. So, not much of a 'cure' then! But surely Peter Daszak would have known about this since-discontinued SARS-CoV vaccine research? Or, was that what was informing him that vaccines wouldn't actually work?

But there is one other option to consider. Let's assume for a minute that Peter Daszak *was* telling the truth in 2016 – or at least he was sharing what he *believed* to be true about SARS-like manipulated coronaviruses; that they had no effective protection against them, as evidenced in the fact that some 50+ years after its discovery, the medical establishment still hasn't come up with a cure for the common cold – which of course, is just another coronavirus. This opens up another door of possibility that actually begins to make some sense. That of the Covid Cabal being fully aware of the existence of this gain-of-function research (which we know they were) and, knowing they had no cure, but equally confident that the knee-jerk reaction of the public to any purported release of any new pathogen would be to demand an antidote, a solution or a vaccine 'asap'; that Big Pharma conspired in advance to have billions of 'injectibles' at the ready – ready to profit enormously at the announcement of any such viral release – even if any such injectibles were completely worthless as a preventable.

If we think about it for a moment, such a scenario would make solid business sense would it not? Not very *moral* sense of course, but immanently profitable nonetheless – especially in context of Big Pharma's immunity from lawsuits. You see, the Covid Cabal already knows (because world expert Dr Daszak inadvertently let it slip in 2016) that no vaccine is going to actually 'work' against such a virus, but if the public doesn't know that, then why ignore such a golden opportunity to make billions out of a fearful, trusting public? Better still, if we can use the fear of a deadly viral release to get the whole world lined up for unproven jabs. Maybe there's another opportunity staring us in the face here? Instead of injecting people with what would be no more than a very expensive placebo, let's see if there's anything else that can be done with such an opportunity while the world rolls up its sleeves in meek compliance? Now that, I suggest, is a far more likely scenario that aligns with the fact that many of the Covid-19 drivers are sociopaths and profiteers with no concern for the damage caused as they scramble for more and more.

THE VACCINES

This brings us to the injections themselves, euphemistically called 'vaccines' and what exactly is contained in them. Because, contrary to popular opinion – even amongst the anti-vaxxers – these 'vaccines' have in fact been very well tested. Not in the standard fashion as we might think as prospective antidotes to SARS-CoV-2 (whatever SARS-2 may yet prove to be) because there simply hasn't been enough time for that, but elements of these particular injectibles – and very similar others – have indeed been 'robustly tested' on animals for years now because first of all there are scores of patents already registered on these purported 'novel' SARS coronaviruses and secondly, because scientific reports confirm that many of those animal tests resulted in serious adverse reactions, multiple organ failures and even the deaths of the test animals.

This was confirmed in a sitting of the Texas Senate Committee of State Affairs on May 6th 2021 where, in a short exchange between Republican Senator Bob Hall and paediatrician Dr. Angelina Farella who was testifying before the Committee, the safety of the Covid-19 vaccines was raised. We repeat that exchange verbatim as lifted from a prominent fact-checking website, in response to the avalanche of Cabal-sponsored rebuttals which try to assert that either the video is a fake; that the Texas hearing didn't happen; and/or even if it did (they say) that they were not discussing Covid vaccines anyway. Which is it? Please decide for yourself: [lxv]

> *Farella: "I've given tens of thousands of vaccines in my office. I am not an anti-vaxxer. I'm very pro-vax actually — except when it comes to this COVID vaccine, if we can call it that."*
>
> *Hall: "Have you seen any other vaccine that was put out for the public that skipped the animal test?"*
>
> *Farella: "Never before, especially for children."*
>
> *Hall: "What I've read, they actually started the animal test, and because the animals were dying, they stopped the test."*
>
> *Farella: "Correct."*
>
> *Hall: "This is important Folks. What we're talking about is that the*

American people are being used as guinea pigs."

We do know that Pfizer, Moderna and Johnson & Johnson confirmed that they had launched Covid-19 vaccine tests on 100,000 volunteers concurrently with animal trials in 2020, but that doesn't explain away the usual 7–10 years of animal testing that is usually required for new medicines like this. What it does explain unfortunately – (that's if we take what they're telling us on face value) – is that Pfizer, Moderna and J & J had no difficulty getting 100,000 people to line up for an experimental procedure that, given the track record of Big Pharma to date, could very well have injured or killed them. One has to wonder if ignorance or credulity were amongst the preconditions for volunteers to step forwards? Because this doesn't say too much about the decision-making qualities of the participants, does it? Ah well, perhaps we should remember that the whole point of using guinea pigs in the first place is to test out potentially-harmful medicines and injectibles on test targets who haven't been properly informed of the dangers, and, even if they have, dumb guinea pigs (or whatever other meek species is chosen) are unlikely to object to the procedure anyway.

So, the cleverly-crafted statement that these 'vaccines' have been 'fully tested in trials' is actually factually correct. But the tack-on declaration (or assumption) that they are therefore 'safe' for human use is not. These injectibles are indeed 'effective' in other words at whatever it is they are designed for – just NOT effective in making us well, and certainly not effective in reassuring us that they are either safe or indeed clinically effective as an inoculation against that indeterminable 'something' which they are labelling Covid-19!

Given the emerging data about all of the 'adverse reactions' being experienced by those who have already been injected, it is probably sufficient to just quote the pertinent part-contents from the official fact-sheets that detail the authorisations and the risks for three of the lead injectibles that have been authorised – sort of – under emergency legislation by the US authorities and endorsed by the WHO; so as to raise the simple but poignant question as to why on earth anyone would want to put this stuff into their bodies?

According to the latest updates on Wikipedia, *"National regulatory authorities have granted Emergency Use Authorizations (EUA's) for twenty-one COVID-19 vaccines. Six of those have been approved for 'emergency' or full use by at least one WHO-recognized regulatory authority (Oxford–AstraZeneca, Pfizer-BioNTech, Sinopharm-BBIBP, Moderna, Sinovac and Janssen)."*[lxvi] Naturally, we're not going to devote half the book exploring each and every vaccine, especially while a massive question remains about what exactly it is that they are creating these vaccines for when we still don't have any truly-and-properly-isolated samples of any supposed 'original SARS-CoV-2 virus' that was NOT created in a lab, and which may yet prove to be no more than a theoretical concept – as we will soon see. So, we are just going to look at the three leading jabs and at the fact-sheets that must by law, accompany any injectibles to see if the Moderna, Pfizer and Janssen data may assist in the investigation. Official sources inform us of the traditional concept of 'vaccination' thus:

> *"Many vaccines work by introducing a weakened or inactive virus or bacteria into the body which triggers the immune system to produce antibodies in response. These antibodies then protect the body if it comes into contact with the real thing".*

Remembering that the common understanding of 'a vaccine' is, "something that is injected into the host that contains either dead-or-living samples of the disease—or transferred antibodies—so as to stimulate the creation of protective antibodies in the host, and to prevent retransmission"; then these three injectibles do not actually qualify for that definition other than by the medical establishment persisting in labelling what is in effect 'a gene therapy treatment' (at best) or more accurately, 'a gene alteration process' – as "a vaccination". But why would they do this? Well, by the time you've read through the next few pages, the fog should have lifted.

> *"Covid-19 mRNA vaccines go one step back in the process. Once inside the body, the mRNA works to build the spike proteins* [that

appear] *on the surface of the Covid-19 virus* [itself]. *The body then responds by producing antibodies which attack those proteins. This means that if it is later infected, the body will be able to generate a faster and more effective immune response to the virus".* [lxvii]

So, the body is being instructed to actually *make* the dangerous 'docking' part of the virus using this new nanotechnology in a process which has of course been robustly tested and isn't at all dangerous - right? Interestingly, all three vaccine fact-sheets contain almost identical warnings, cautions, advisories and preambles, so for ease of understanding we have generated a combined list and replaced the names 'Moderna, Pfizer & Janssen' with the collective term 'The Vaccine' noting the exceptions where appropriate. Underlines have been added for emphasis and text in *italics* appears to be untrue or misleading statements that align with the official narrative. Text [in square brackets] has been added by the author to assist in clarity.

THE MODERNA, PFIZER AND JANSSEN FACT-SHEETS: The [named] Vaccine is a vaccine and <u>may</u> prevent you from getting COVID-19. <u>There is no U.S. Food and Drug Administration (FDA) approved vaccine to prevent COVID-19</u>. It is your choice to receive the Vaccine. The Vaccine <u>may not protect everyone</u>. COVID-19 is caused by a coronavirus called SARS-CoV-2. *This type of coronavirus has not been seen before.* You can get COVID-19 through contact with another person who has the virus. It is predominantly a respiratory illness that can affect other organs. People with COVID-19 have had a wide range of symptoms reported, ranging from mild symptoms to severe illness. Symptoms may appear 2 to 14 days after exposure to the virus. Symptoms may include: fever or chills; cough; shortness of breath; fatigue; muscle or body aches; headache; new loss of taste or smell; sore throat; congestion or runny nose; nausea or vomiting; diarrhea. <u>The FDA has authorized the emergency use of the Vaccine to prevent COVID-19 in individuals [12 or] 18 years of age and older under an Emergency Use Authorization (EUA)</u>.

Tell your vaccination provider about <u>all of your</u> [existing]<u> medical conditions</u>, including if you:

> **[All 3 Vaccines]** • have any allergies • have had myocarditis (inflammation of the heart muscle) or pericarditis (inflammation of the lining outside the heart) • have a fever • have a bleeding disorder or are on a blood thinner • are immunocompromised or are on a medicine that affects your immune system • are pregnant or plan to become pregnant • are breastfeeding • have received another COVID-19 vaccine.

> [plus] **PFIZER:** • have ever fainted in association with an injection.

> <u>You should not get the Vaccine if you</u>: • had a severe allergic reaction after a previous dose of this vaccine • had a severe allergic reaction to any ingredient of this vaccine.

<u>The Vaccine is an unapproved vaccine</u>. *In an ongoing clinical trial, the Vaccine has been shown to prevent COVID-19.* <u>The duration of protection against COVID-19 is currently unknown.</u> There is a *remote* chance that the Vaccine could cause a severe allergic reaction. <u>A severe allergic reaction would usually occur within a few minutes to one hour after getting a dose of the Vaccine</u>.

Signs of a severe allergic reaction can include:

> **[All 3 Vaccines]** • Difficulty breathing • Swelling of your face and throat • A fast heartbeat • A bad rash all over your body • Dizziness and weakness Myocarditis (inflammation of the heart muscle) and pericarditis (inflammation of the lining outside the heart) have occurred in some people who have received the Vaccine. In most of these people, symptoms began within a few days following receipt of the second dose of the Moderna & Pfizer Vaccines and within 1 – 2 weeks of the Janssen Vaccine. The chance of having this occur is *very low*. Side effects that have been reported in clinical trials with the Moderna & Pfizer Vaccines include: • Injection site reactions: pain, tenderness and swelling of the lymph nodes in the same arm of the injection, swelling

(hardness), and redness • General side effects: fatigue, headache, muscle pain, joint pain, chills, nausea and vomiting, and fever.

[plus] **PFIZER:** Side effects include: • severe allergic reactions • non-severe allergic reactions such as rash, itching, hives, or swelling of the face • myocarditis (inflammation of the heart muscle) • pericarditis (inflammation of the lining outside the heart) • injection site pain • tiredness • headache • muscle pain • chills • joint pain • fever • injection site swelling • injection site redness • nausea • feeling unwell • swollen lymph nodes (lymphadenopathy) • diarrhea • vomiting • arm pain. These may not be all the possible side effects of the Pfizer-BioNTech COVID-19 Vaccine. Serious and unexpected side effects may occur.

[plus] **JANSSEN:** • Difficulty breathing, • Swelling of your face and throat, • A fast heartbeat, • A bad rash all over your body, • Dizziness and weakness. Blood Clots with Low Levels of Platelets Blood clots involving blood vessels in the brain, lungs, abdomen, and legs along with low levels of platelets (blood cells that help your body stop bleeding), have occurred in some people who have received the Janssen COVID-19 Vaccine. In people who developed these blood clots and low levels of platelets, symptoms began approximately one to two-weeks following vaccination. Reporting of these blood clots and low levels of platelets has been highest in females ages 18 through 49 years. The chance of having this occur is *remote*. You should seek medical attention right away if you have any of the following symptoms after receiving Janssen COVID-19 Vaccine: • Shortness of breath, • Chest pain, • Leg swelling, • Persistent abdominal pain, • Severe or persistent headaches or blurred vision, • Easy bruising or tiny blood spots under the skin beyond the site of the injection. These may not be all the possible side effects of the Janssen COVID-19 Vaccine. Serious and unexpected effects may occur. Side effects that have been reported with the Janssen COVID-19 Vaccine include: • Injection site reactions: pain, redness of the skin and swelling. • General side effects: headache, feeling very tired, muscle aches,

nausea, and fever. The Janssen COVID-19 Vaccine is still being studied in clinical trials. Guillain Barré Syndrome Guillain Barré syndrome (a neurological disorder in which the body's immune system damages nerve cells, causing muscle weakness and sometimes paralysis) has occurred in some people who have received the Janssen COVID-19 Vaccine. In most of these people, symptoms began within 42 days following receipt of the Janssen COVID-19 Vaccine. The chance of having this occur is *very low*. You should seek medical attention right away if you develop any of the following symptoms after receiving the Janssen COVID-19 Vaccine: • Weakness or tingling sensations, especially in the legs or arms, that's worsening and spreading to other parts of the body • Difficulty walking • Difficulty with facial movements, including speaking, chewing, or swallowing • Double vision or inability to move eyes • Difficulty with bladder control or bowel function

If you experience a severe allergic reaction, call 9-1-1, or go to the nearest hospital. V-safe is a new voluntary smartphone-based tool that uses text messaging and web surveys to check in with people who have been vaccinated to identify potential side effects after COVID-19 vaccination. V-safe asks questions that help CDC monitor the safety of COVID-19 vaccines. V-safe also provides second-dose reminders if needed and live telephone follow-up if participants report a significant health impact following COVID-19 vaccination. It is your choice to receive or not receive the Vaccine. [This 'choice' seems to be diminishing fast.] Currently, there is no FDA-approved alternative vaccine available for prevention of COVID-19. Other vaccines to prevent COVID-19 may be available under Emergency Use Authorization [EUA]. There is no information on the use of the Vaccine with other vaccines. If you are pregnant or breastfeeding, discuss your options with your healthcare provider.

The Vaccine does not contain SARS-CoV-2 and cannot give you COVID-19. [So why is it being called a 'vaccine'?]

The vaccination provider *may* include your vaccination information in your state/local jurisdiction's Immunization Information System (IIS) or other designated system. At this time, the provider cannot charge you for a vaccine dose and you cannot be charged an out-of-pocket vaccine administration fee or any other fee if only receiving a COVID-19 vaccination. However, vaccination providers may seek appropriate reimbursement from a program or plan that covers COVID-19 vaccine administration fees for the vaccine recipient (private insurance, Medicare, Medicaid, HRSA COVID-19 Uninsured Program for non-insured recipients). Individuals becoming aware of any potential violations of the CDC COVID-19 Vaccination Program requirements are encouraged to report them to the Office of the Inspector General.

The Countermeasures Injury Compensation Program (CICP) is a federal program that *may* help pay for costs of medical care and other specific expenses of certain people who have been seriously injured by certain medicines or vaccines, including this vaccine. The United States FDA has made the Vaccine available under an emergency access mechanism called an EUA. The EUA is supported by a Secretary of Health and Human Services (HHS) declaration that *circumstances exist to justify the emergency use of drugs and biological products during the COVID-19 pandemic.*

The Vaccine has not undergone the same type of review as an FDA-approved or cleared product. FDA may issue an EUA when certain criteria are met, which includes that *there are no adequate, approved, and available alternatives.* In addition, the FDA decision is based on the totality of the scientific evidence available showing that the product may be effective to prevent COVID-19 during the COVID-19 pandemic and that *the known and potential benefits of the product outweigh the known and potential risks of the product.* All of these criteria must be met to allow for the product to be used during the COVID-19 pandemic. The EUA for the Vaccine is in effect for the duration of the COVID-19 EUA declaration justifying emergency use of these products, unless terminated or revoked (after which the products may no longer be used).

The Moderna COVID-19 Vaccine [double dose, 18 years and older, 1 month apart] <u>contains</u> the following ingredients: <u>messenger ribonucleic acid (mRNA)</u>, lipids (SM-102, polyethylene glycol [PEG] 2000 dimyristoyl glycerol [DMG], cholesterol, & 1,2-distearoyl-sn-glycero-3-phosphocholine [DSPC]), tromethamine, tromethamine hydrochloride, acetic acid, sodium acetate trihydrate, & sucrose.

The Pfizer-BioNTech COVID-19 Vaccine [double dose 3 weeks apart, 12 years and older] <u>includes</u> the following ingredients: mRNA, lipids ((4-hydroxybutyl)azanediyl)bis(hexane-6,1-diyl)bis(2-hexyldeca-noate), 2 [(polyethylene glycol)-2000]-N,N-ditetradecyl-acetamide, 1,2-Distearoyl-sn-glycero-3-phosphocholine, & cholesterol), potassium chloride, monobasic potassium phosphate, sodium chloride, dibasic sodium phosphate dihydrate, & sucrose.[lxviii]

The Janssen COVID-19 Vaccine [single dose, 18 years and older] <u>includes</u> the following ingredients: <u>recombinant</u>, replication-incompetent adenovirus type 26 *expressing the SARS-CoV-2 spike protein*, citric acid monohydrate, trisodium citrate dihydrate, ethanol, 2-hydroxypropyl-beta-cyclodextrin (HBCD), polysorbate-80, sodium chloride.[lxix]

So is that all of it? Could we have possibly missed anything here? Well, on the WHO's official data on the Janssen Vaccine we read that sodium hydroxide and hydrochloric acid are also ingredients. So why aren't they included in the list above? Maybe that's the reason for the flexible term 'includes' vs 'contains'? It should also be clarified that the Moderna and Pfizer vaccines are using mRNA technology (more on this in a moment) whilst the Janssen jab – somewhat misleadingly (if we are to believe the "does not contain SARS-CoV-2" statement) is using a combination of altered SARS-CoV-2 elements along with extracts from cultured aborted foetal tissue to convey the vaccine to its intended destination. Maybe scientifically necessary to the task perhaps, but nevertheless morally disturbing. In fact, and speaking purely for myself, as soon as I heard that aborted foetal tissue was included in some of these vaccines, and used in the development of nearly all of them, all of my internal alarms went off.

A couple of obvious issues immediately arise with the data we just read. Each company warns that you should NOT take their vaccine, *"..if you have received another COVID-19 vaccine"* and that, *"there is no information on the use of the Vaccine with other vaccines"*. So why, on the news today do we hear people being advised to return to the vaccination centres to receive top-up doses of competing vaccines?

Another issue is the recommended age limits of either 'over 12' or 'over 18'. How is it – even with all the emerging evidence of massive, wholesale adverse reactions pouring in, that moves are now afoot to inject babies and children, with those in State care not even having a personal choice in the matter? In fact, it has since become clear that 'the State' – in various locations around the world – now plans to bypass parental consent and just go ahead and inject our children with whatever they see fit whenever they see fit; tested or not; safe or not; or medically necessary or not, in a move that is the very definition of imperious fascism. The fact that so many State officials now know that these injectibles are dangerous takes this planned 'initiative' to another and more sinister level, does it not? What price the lives of our children Folks? What price our silence and our compliance with the sociopaths and with the diabolicals?

In the process of writing this book, a dear friend of many years – an old soldier in a prestigious military care home in the UK – informed me that he had finally succumbed to official threats that if he didn't take the jab he would be expelled from the facility. They even laid on a special ambulance to transport him to an outside clinic after he had successfully refused multiple attempts for in-house injections. Shortly after receiving the Pfizer/BioNTech jab, John woke up completely and permanently blind in his left eye. He is now being asked to sign a retroactive disclaimer...

Some of the jab contents and their functions in layperson's terms:

mRNA	Hijacks your cell's ribosomes (internal protein factories) to start making spike proteins inside your body that mimic those of SARS-CoV-2
SM-102	Helps shield the mRNA from attack by the body's natural immune system en route to cells
PEG	Petroleum-based compound used as a preservative
HCG	Infertility chemical and spontaneous abortion trigger
DMG	Athletic performance enhancer (plus more)..
DSPC	Drug delivery vehicle in these vaccines
HBCD	Highly toxic flame retardant, on the 'substances of very high concern' list and banned in Japan since 2014, and tightly controlled by the EU since 2015
acetic acid	Vinegar-like descaling agent
phosphocholine	An immune-response suppressive found in human placentas and chicken eggs
potassium chloride	Used in fertilisers and farming, as well as lethal injections due to its ability to cause cardiac arrest
recombinant	An artificial DNA sequence
ethanol	Alcohol and psychoactive drug
polysorbate-80	Emulsifier and excipient
sodium hydroxide	Highly caustic compound that dissolves proteins and body tissues
hydrochloric acid	Strong inorganic acid used in the treatment of metals

Beginning at the top here, the great, and very well-known risk of the mRNA process is that it can create an autoimmune response in the host which basically means that the body starts fighting itself trying to get rid of the 'foreign stuff' (the S1 'docking' spike proteins that appear on the outside of a coronavirus) – the very same spike proteins which the body's own cells are now producing!? The potential 'suicidal collapse' of the immune system in other words, in what is known as a cytokine storm. And worse still, this reaction can occur many years after the injection believe it or not! In fact, as we discuss later in the book, vaccine research into SARS-CoV-1 in 2002-03 established the existence of this Antibody Enhancement Disease (ADE) before that particular research was discontinued; noting that whatever protection was being delivered to the test animals was being cancelled out by these ADE autoimmune reactions resulting in death. The same data tracks back to early coronavirus experiments in the 1960's where a number of children actually died as a result.

It seems obvious to me, and as reported by eminent credible experts, that if these S1 spike proteins are the very things that help viruses infect human cells, then whose brilliant idea was it to introduce these dangerous elements right inside human cells where, once it gets in, an intelligent little virus like SARS-CoV-2 can make immediate use of all of these ready-to-rock specialised viral appendages that the host has now very conveniently placed at the virus' disposal? Like some sort of Trojan Horse scenario that could very well be putting the enemy right where it can do the most damage? Sort of like armies and guns, with a complete coronavirus representing an armed platoon, and the spike proteins being represented by the guns. Why would our own generals (the biotech scientists) send crates of enemy guns (the spike proteins) right inside our own defences? Because if for instance we (our bodies) don't properly understand how to use those guns in our own defence, then the whole experiment could very easily backfire, right? Something that does now seem to be happening with all of the adverse reactions and vaccine-related deaths being reported.

Why indeed would the authorities be taking such risks with an experimental procedure when they could just as easily inject people with a piece of inactivated (dead) SARS-CoV-2 virus as has been done with the traditional (and ever-changing) annual flu shot? After all, if it works okay for the flu (which is a similar virus with parallel symptoms) and has done so in a tried-and-tested fashion albeit only around 10-60% effective for decades but *without* causing ADE, then why are they departing from that science now to experiment (at very best) with these speculative and potentially very dangerous Trojan Horse-type procedures? Alternatively, why haven't they used the other standard MMR vaccination technique; that of injecting a very weakened 'attenuated' version of the virus to stimulate antibodies *without* causing the host to get too sick? If the intention is to create protective antibodies inside the injected host, then we already have tried-and-tested methods – so why aren't they doing that instead?

May I suggest that it's either because; (a) they don't actually have a piece of genuinely isolated live-or-dead SARS-CoV-2 virus in the first place to do this, and/or; (b) because the use of mRNA materials is already written into the script and has been prepared long in advance as a proposed 'effective response' or as the necessary 'next step' and, either out of scientific curiosity or diabolical intent, that's what's going to be injected into us whether we like it or not! The fact that stock market prices peaked at exactly the same time at the announcement that Big Pharma had somehow made this miraculous vaccine in record time – complete with abundant evidence of insider trading-type speculations and anticipations, surely seals the deal?

In fact, it's probably time to call these mRNA injectibles exactly what they are. They are in fact a proposed untested 'treatment' for Covid-19 that have used 'assays' (diagnostic test elements) from existing coronaviruses including from SARS-CoV No.1, to create an internal reaction inside the human body. The pushers have misleadingly named these experimental treatments 'vaccines' so as to benefit from the legislation that allows for emergency use authorisations (EUA). They also wish to avoid the awkward conversation as to what *other* treatments could be used instead to treat this purported illness, such as ivermectin, hydroxychloroquine, or chicken soup, vitamin C and a day in bed? As it is, the mRNA injections are causing a reaction inside the cells of the human body that are responding by actually *creating* the pathogenic elements of the coronavirus (spike proteins/guns) that are then somehow meant to provide protection for the host!? But this doesn't make sense, because first of all they are claiming that some 80% of the people who are 'infected' with Covid are not symptomatic – meaning they have no symptoms at all – but somehow, after being injected, that 80% then develop adverse reactions to the vaccines?? This clearly suggests that the real physical problem that is causing sickness here is not an invisible, non-symptomatic virus (which may not even exist) but is the artificially-constructed treatments for 'it' that are delivering pathogens directly into human cells!? The intriguing fact that one of the coronavirus-drawn testing assays used to construct the vaccines is named 'luciferase' may have more than a coincidental symbolic reference to what may yet prove to be a diabolical experiment with preplanned intentional outcomes.[lxx]

*   *   *

Before we go any further, perhaps this is an opportune moment to quote from the CDC's own database about many of the standard elements that go into traditional vaccines. Vaccines which Dr Vernon Coleman and many, many others keep on telling us are not at all safe, raising the uncomfortable question again of whether it could be possible that the genuine success of early vaccination programs against diseases such as anthrax, ebola, measles, mumps and polio for instance has been hijacked by the pharmaceutical industry and paired up with gain-of-function research in a never-ending cycle of profit-generating diseases, leading to profit-generating research into profit-generating vaccines? Or indeed, by using the well-accepted concept of 'necessary vaccinations' for any purpose that might fit with the globalist agenda, such as finding some way to get 'stuff' into our bodies that will service the futuristic control ambitions of the Cabal in the form of information-gathering nanoparticles, sterilising agents or even mood reactors?

Some of those typical vaccine ingredients include: [lxxi]

- Formaldehyde/Formalin; a highly toxic poison and carcinogen.
- Betapropiolactone; a toxic chemical and carcinogen.
- Hexadecyltrimethylammonium bromide; can cause damage to the liver, cardiovascular system, and central nervous system, as well as cause reproductive issues and birth defects.
- Aluminium hydroxide, aluminium phosphate, and aluminium salts; all neurotoxins carrying long-term risks of brain inflammation and swelling, causing neurological disorders, autoimmune disease, Alzheimer's, dementia, and autism.
- Thimerosal (mercury); another neurotoxin. Induces cellular damage, reduces oxidation, causes cellular degeneration, and cell death. Also linked to neurological disorders, Alzheimer's, dementia, and autism.
- Polysorbate 80 & 20; can penetrate the brain membrane carrying aluminium, thimerosal, and viruses.
- Glutaraldehyde; another toxic chemical used as a disinfectant.
- Foetal Bovine Serum; harvested from bovine (cow) foetuses taken from pregnant cows before slaughter.
- Human Diploid Fibroblast Cells; aborted foetal cells from babies
- African Green Monkey Kidney Cells; these may carry the SV-40 cancer-causing virus that has affected at least 30 million Americans.
- Acetone; that can cause kidney, liver, and nerve damage.
- E.Coli; being promoted as a delivery vehicle for new vaccines.
- DNA from porcine (pig) Circovirus type-1. [lxxii]

As we keep saying; who in their right mind would actually volunteer to have any of this 'stuff' injected into them? Notwithstanding all of the adverse reactions, and chronic illnesses and deaths that have been caused by the reckless profiteering of Big Pharma who, as one would expect, remain fully indemnified by our governments for any liability for damage caused by the emergency rollout of Covid-vaccines, is it really any surprise to hear that more people have now died from the Covid-19 vaccines than all of the casualties from ALL of the vaccines given in the USA (for example) in the last twenty years combined – fast approaching 30 years![lxxiii] And this, even when considering the statistical likelihood – according to a recent study by Harvard University in conjunction with the CDC's own VAERS[10] agency – that we are only actually hearing about 1% of those casualties!? [lxxiv]

Autism in children is another vaccine-related example. Comprising only about 1 case in 10,000 in 1999 when various 'National Vaccine Initiatives' worldwide promised to, "stamp out childhood diseases". How then have we arrived at a scenario where the incidence of autism in children who have been traditionally vaccinated now stands at 1 in 32? Personally, I would like to see all vaccines being publically tested in fully-documented clinical trials on the Executives and CEO's of Big Pharma as well as the biotech scientists and Wall Street profiteers before they ever dare to try to market them to the public.

In fact, even as certain prominent politicians such as Angela Merkel and Boris Johnson posed for the cameras apparently getting their Covid-19 shots – but for some strange reason, there weren't any visible needles as the air-filled plunger was depressed by a fawning nurse into Mrs Merkel's upper arm; nor did we get anything more that a highly suspicious still shot (in the middle of a typically-garrulous Boris Johnson video?) of a pretty British nurse bending over Mr Johnson promisingly with a loaded syringe; oblivious apparently to the real danger she was actually in. The question as to how and why a reportedly "previously infected" Boris Johnson has not acquired Covid immunity was simply not asked. Why ever not?

---

[10] VAERS: Vaccine Adverse Effects Reporting System.

Are the experts who are grouping 'the vaccinated' and 'the immune' together, wrong again? One wonders then, as all these world leaders creep towards the seemingly-inevitable "mandatory jabs" declaration, why, with the majority of American businesses also mandating that their employees either get vaccinated or get their employment contracts terminated – yet US Pfizer employees are under no such obligations? Does Pfizer perhaps know something that we don't? After all, they're still going to need employees – right? [lxxv]

Nobel Laureate Professor Luc Montagnier is another world-leading virologist. The fact that discredited fact-checkers are now hot on his trail prompts me to quote exactly what he said as he described any planned mass-vaccination program during a pandemic as;

"Unthinkable! It's a historic blunder and an unacceptable mistake that is creating the variants and leading to deaths from the disease."

Speaking of Antibody Dependant Enhancement disease (ADE) Dr Montagnier makes the unequivocal statement that, *"In every country the curve of vaccinations is followed by a curve of deaths."* He further maintains that a healthy immune system is enough against most common viruses (Big Pharma didn't like that one too much either) and he was also one of the first to state that SARS-CoV-2 was not a naturally-occurring phenomenon, but had been artificially engineered in a lab – probably in Wuhan, China. And yes, the fact-checkers did their best to debunk him then too. Other world-renowned doctors including Dr. Peter McCullough of Texas A & M College of Medicine and vaccine creator Geert Vanden Bosshe PhD (also of the B & M Gates Foundation) have been warning anyone who will listen that this was never about any genuine health crisis but that it is all about, "getting needles in every arm" despite the death rate of Covid vaccines being much higher than any other vaccine – or indeed combination of vaccines – in history. That Big Pharma, the WHO, the CDC, the NIH, the Gates Foundation and their political backers are all pushing a horribly dangerous process that is not only killing and disabling people in the short term, but that will result in compromised immune systems that will leave the vaccinated in particular, hugely vulnerable even to mild pathogens.

But at least the people who have taken the Covid vaccines are safe now from infection, re-infection or transmission – right? Nope, wrong again! Because they're NOT actually 'vaccines' as doctors understand the term, as is now increasingly clear. Adding another sinister component to the mix is the prospect that these new injectibles – based upon the anticipated Covid-19 variants coming down the tracks – which variants will be named after the Greek alphabet and thereafter after constellations of stars (guess how many of those there are); that these mandatory jabs can and will be used as a means of delivering microscopic tracking gels and other information-gathering materials that will respond to sensors and satellites thereby effectively monitoring everything that we do. And 'monitoring' of course is just one short step away from 'controlling'. And yes! This type of nanotechnology does already exist as is being used in military surveillance for example and can be seen in the June 2019 patent application registered by Bill Gate's Microsoft Corporation for, 'A CRYPTOCURRENCY SYSTEM USING BODY ACTIVITY DATA' which is registered, very interestingly under patent application number WO2020-060606. [lxxvi]

*   *   *

Just as we were completing this book, a colleague alerted us that despite the Janssen jab being withdrawn from circulation in Ireland in early August due to multiple reports of blood-clotting, that the Health Service Executive (HSE) was facilitating the continued use of the vaccine by unscrupulous doctors and pharmacists advising them to backdate the date of injection to before the withdrawal date. This is to ensure that everyone in the loop gets paid apparently. Yet another example of what happens when profit comes before people, and when we trust establishment figures with our health![lxxvii]

Mention is made of the Nuremberg 2 Trials which many of us are hoping will come sooner, rather than later.

CHAPTER SEVENTEEN

CLASSIFICATIONS & CURES

A SHORT HISTORY OF CORONAVIRUSES

Doing our best to keep the scientific jargon to a minimum, it will probably help us all to get our heads around the physical and microbiological aspects of the whole Covid-19 phenomenon with a short but interesting foray into the history of coronaviruses to find out what exactly they are and where they really came from, as well as looking at the controversy over the inexplicable banning of proven viral cures such as ivermectin and hydroxychloroquine. First, a reminder of the main culprits:

- **HCoV-229E:** discovered in 1965 in Chicago. A common cold with generally mild symptoms.

- **HCoV-OC43:** discovered in 1969, as documented in outbreaks since 2003, also a common cold with mild symptoms.

- **SARS-CoV-1:** Severe Acute Respiratory Syndrome coronavirus detected in Asia in late 2002. Can cause 12% deaths.

- **HCoV-NL63:** discovered in 2004 in the Netherlands. Another respiratory infection with generally mild symptoms.

- **HCoV-HKU1:** discovered in Hong Kong in 2005. Also resulting in generally mild symptoms.

- **MERS-CoV:** Middle East Respiratory Syndrome-related coronavirus occurred in 2012 & 2018 in Saudi Arabia and in South Korea in 2015. Highly dangerous to some. 35% deaths.

- **SARS-CoV-2:** Severe Acute Respiratory Syndrome coronavirus 2 – the truth about which we hope to clarify in this book.

We'll begin by reiterating that coronaviruses – and especially the small handful that affect humans – are zoonotic – meaning that they originate in animals and are then believed to trans-migrate through various mutations into human hosts. They are also 'recombinant' organisms, meaning that they combine when two (or more) viral genomes (the original materials from two different pathogens) are present in the same host cell.[lxxviii] They can 'combine' in other words inside an animal's body to become a new strain, or variant if you like, which, because it is now different than either of the two original pathogens, can go on to infect new and different species – or hosts. We are told that these recombinants can arise either by natural occurrence or by human intervention by highly-skilled virologists and scientists – something we already know is possible because of all of the 'recombinant' patents already on the record. In short, we know for a fact that human-infecting coronaviruses can definitely be 'created' in the lab, but it's *still* only the theoretical conjecture of scientists that these organisms have in fact been around for centuries, 'waiting to be discovered'. If the reader does a quick online check of the standard sources of information (quickly, before it all gets fact-checked and suitably 'corrected') you will see that explanations of the origins of human-infecting coronaviruses are all prefixed with terms such as, "We believe.." or "It is assumed.." or "Scientists think that.." etc., so there is no *definitive* scientific consensus as to the empirical origins of coronaviruses other than what they *think* may have happened. This raises a very serious question as to whether or not human-infecting coronaviruses could, possibly, have been the result of human interference and research into pre-existing animal viruses, and whether it was that very interference that then made them into zoonotic viruses that then went on to infect and kill people?

Now I know that this idea is going to be rebuffed and dismissed as an outrageous suggestion by the rebutters and fact-checkers – and probably by most of those who have not done the research; but please understand that this altogether-not-implausible conclusion has only been reached based upon the known facts and the historical record, where before we go any further, we need to note the crucial fact that of each of the seven human-affecting coronaviruses, that all but two of them were 'discovered' or 'first documented' *after* the arrival of the SARS-CoV-1 pandemic of 2002-03, and one of those two exceptions (HCov-OC43) was not even documented as 'an outbreak' of the common cold until 2003 and then not sequenced until 2012.

Given that we know for a fact that several coronavirus-related patents that were explicit to the unique properties of SARS-CoV-1 had been registered *in advance* of that 2003 date – as well as for the methods of manipulating and altering existing coronaviruses to effectively 'create' a SARS-CoV pathogen – thereby proving foreknowledge of SARS's existence by the patent holders, at least in theory, *before* it became public knowledge via the 2003 pandemic; then surely the question of SARS' real origins must be considered – as well as that of the other latter-dated coronaviruses, especially in light of all of the foundations laid out in Part One, where there can be no doubt that an amoral Cabal comprising conscience-less, profit-driven sociopaths and psychopaths has long since been running the world?

But that still leaves us with HCoV-229E, the 1st common cold virus that was reportedly discovered in Chicago in 1965, along with the apparently 'accidental' discovery of the very similar HCoV-OC43 shortly afterwards by a team of scientists at the USA's National Institute of Health (NIH) and the corresponding National Institute of Allergies and Infectious Diseases (NIAID) which is now run by Dr Anthony Fauchi. How do we explain those events in context of a suspicion that those coronaviruses may have been man-made constructions too, especially considering the claim by Forbes Magazine that HCoV-229E wasn't even fully sequenced until 2012? How does this fit with the contradictory assertion that, "The scientific community has been actively – but so far unsuccessfully – seeking a cure for the common cold" (HCoV-229E & HCoV-OC43) "..for decades"? How could they possibly be doing this without having sequenced the OC43 virus itself? Again, in a phrase that keeps repeating itself throughout this book, "something's not right here!"

It is simple commonsense is it not, that unless you understand the problem you're trying to tackle, then you're unlikely to be able to produce a solution. Concept, plan, action – right? Or, (i) first we (scientists) know about the existence of a virus, then (ii) we study its properties and understand how it works, and (iii) we start working on a cure. So how can the scientific community claim they were working on vaccines to tackle an identified pathogen from 1969 which they didn't even bother to fully sequence until 2012? They could of course have been diligently (but apparently unsuccessfully) working away on the other, first-discovered HCoV-229E virus I suppose, but given the massive profits to be made from finding a jab-cure for the common cold, it seems ridiculously implausible that the Big Boys (and Girls) like Fauchi, Baric, Daszak, Zhengli, Gerberding and their virologist colleagues – plus all of the Bill Gates-connected agencies and corporations with vested interests in anything medicine-and-profit related – that they would just sit idly by in knowledge of the existence of another common-cold virus for over 40 years, and not even bother to sequence it? How do we explain this apparent disinterest in something so important to global health especially in context of what's actually going on today? Maybe it's because we're not as important as chickens to them? Because even the 1960's viral researchers concede that they were only working on poultry viruses in the first place, because of the global commercial interest.

Let's have a closer look at coronavirus research itself, because again there are some possible smoking guns here. The first being the historical fact that scientists were already heavily invested in animal-based coronavirus research both in the UK and the USA in the 1960's when they 'discovered' (as opposed to possibly 'created' – accidentally or purposefully?) the first human coronaviruses. In considering this unstated possibility, we should not dismiss the likelihood that during the course of their investigative work where they were in effect messing about with avian, bovine and canine coronaviruses in not-too-strict laboratory conditions trying to see how they might control these diseases for commercial gain; then is it not a distinct possibility that someone, somewhere tweaked this-or-that element in just such a manner, or added a pinch of this and a splash of that so as to provide the opportunity for one of these super-tiny little coronaviruses to mutate sufficiently to infect humans?

As to the possibility that this may have been done deliberately? Well, maybe it was, and maybe it wasn't, and maybe, just maybe it all happened quite naturally, but we shouldn't underestimate the curiosity of diligent scientists and the appeal of being the very first person to find or create something novel that the scientific community hasn't seen before. One's 'name up in lights' and all that.

The next possible red flag is the direct involvement of the NIH and the NIAID in the 1960's, and their (arguably amoral) use of human foetal tissue to cultivate the viruses that they were researching.[lxxix] Is it not probable that by culturing these animal viruses in human tissue there was a distinct risk of a mutation? I don't know the answer to that one, but it sounds like another question that at least deserves an answer. Especially in light of the NIH and NIAID's recent track record and the use of human foetal tissue in the generation of these new RNA injectibles?

Then there's the interesting quote by Dr Ken McIntosh of Harvard Medical School at the time of discovery that, *"These viruses share the characteristics of being easily destroyed by ether."* So, why isn't ether listed amongst the components of the Covid-19 vaccines? I am not a virologist of course so I apologise if this is an inane comment, but it does seem to be another question worth answering, especially in light of the global efforts by the establishment to ban and suppress the use of proven anti-pathogenic and viral-suppressing agents such as hydroxychloroquine (an anti-malarial) and ivermectin (an anti-parasitic) as possible generic cures for all things Covid-related in the opening months of the Covid-19 pandemic.[lxxx]

Ivermectin & Hydroxychloroquine. Backing up the preventative and curative aspects of both ivermectin and hydroxychloroquine against various pathogens including viruses and bacteria – not to forget the symptoms of the common cold and the seasonal flu – are historical scholarly articles where ivermectin receives high praise as a viral inhibitor for example, with the News Medical Life Sciences outlet describing how it can kill SARS-CoV-2 in just 48 hours;[lxxxi] and with the prestigious Lancet Journal stating in 2011 that;

> *"Chloroquine has in-vitro activity against influenza and could be an ideal candidate for worldwide prevention of influenza in the period between onset of a pandemic with a virulent influenza strain and the development and widespread dissemination of an effective vaccine."* [lxxxii]

Sounds like it was tailor-made for the job, doesn't it? So why wasn't it used? Various natural remedies too have been downplayed and ridiculed with our silly notions about us boosting our immune systems with zinc, and vitamins C and D, plus maintaining a healthy lifestyle; all being replaced by the official mantra to just, 'shut up and go get the bloody jab!' That whole 'what other options are there' topic needs to be independently researched, because this is a clear indicator that the decision-makers at the WHO, plus officials at the NIH, the CDC, the FDA, and the NIAID in the USA, as well as the European Medicine Agency, the UK's NHS, and Ireland's HSE for example, plus most governments around the world were not at all in fact interested in any cheap, effective and readily-available cures for this Covid-19 pathogen, but wanted instead to herd us all into testing and vaccination centres NOT for the greater good of humanity, but for the greater profit of Big Pharma and to fulfil the hidden agenda.

A quick search of the internet will reveal all manner of flip-flopping and contrary statements-and-positions by the so-called 'health authorities' on ivermectin and hydroxychloroquine in particular, as an increasingly distressed and better-informed public now seeks to understand why the Cabal was so vehemently opposed to using these cheap and readily available drugs in the fight against Covid – so much so that they actually made it unlawful to sell these products in places like the UK, France, Italy, Belgium and Australia with massive fines for those who did.[lxxxiii] But why was the establishment so dead-set against these particular medicines during the early months of the pandemic? What was the problem *then* – as opposed to now – where we see US agencies and even the WHO now tentatively 'approving' ivermectin and chloroquine in strictly-controlled clinical trials in an apparent smoke-and-mirrors operation that is keeping people distracted and confused about the whole Covid-19 phenomenon? [lxxxiv]

Meanwhile, some 40,000 frontline workers in Europe, Africa, Asia and South America, are (we are told) already involved in clinical trials using either chloroquine, hydroxychloroquine or a placebo to test their effects on Covid symptoms. So why the apparent belated about-turn by the Cabal?[lxxxv]

Now we really need to pay attention here and ask a pretty obvious question: what was it about ivermectin and hydroxychloroquine that had the cabal so very worried? And why are they apparently relaxing their objecting position now? Well, eliminating the possibility that this has all been an unplanned 'charlie-foxtrot' fiasco, there are four straightforward answers to these questions which strike at the very heart of the whole Covid-19 phenomenon, and it's really important that the reader sticks with the logic here as we do our best to explain this issue without delving unnecessarily into scientific complexities.

First of all there's the obvious reason for the establishment's stoic objections to ivermectin and hydroxychloroquine; that there was more money to be made out of new tests and vaccines than by availing of existing, available tried-and-tested methods. Not very admirable, but understandable in the circumstances. Unfortunately, 'just making money' doesn't account for all of the other impositions such as social distancing and lockdowns, even though it is clear that the establishment intends profiting from those dim realities as well.

Secondly, that because there was a pre-planned sequence of events from; (i) declaration of a pandemic, to; (ii) masks and social distancing, to; (iii) PCR Testing that was going to pick up on anything coronavirus-related, to; (iv) tracking-and-tracing, to; (v) vaccination... and from there to whatever other draconian measures the establishment planned to impose upon us (passports, exclusion zones, mandatory jabs); that it was imperative that, (a) the public remained in a constant state of fear and apprehension, so that, (b) they would follow the program and complete the (i)–(v) sequence as swiftly and efficiently as possible. Accordingly, introducing possible other cures or preventatives at the very *start* of the sequence, was NOT part of the plan and, if the authorities were to allow us access to those other cures, then we might all stop taking the whole Covid-19 'crisis' so seriously – right? Well, they couldn't let that happen, now could they?

The third possible explanation for the 'safe-drugs ban' revolves around the misuse of the PCR Tests as a viral diagnostic, whereby the Cabal needed these tests to pick up on any-and-all cold-or-flu related symptoms to boost the case numbers, and they needed lots of people to get tested. But if a terrified populace decided to self-treat themselves or seek out honest physicians who were prepared to prescribe preventative drugs like ivermectin and chloroquine, well, that was obviously going to affect the numbers turning up at test centres, and the number of 'positive' (or false-positive) results, whilst also steering the people away from the officially-sanctioned injectible solutions and unquestioned obedience to the diktats of the Cabal. This would also partly explain why honest, dissenting doctors have been so very harshly treated.

The fourth, more sinister explanation requires an understanding of the science that has been uncovered by the discovery of the SARS-related patents – which we will discuss shortly. But in simple terms, the SARS-CoV-2 virus itself has apparently been artificially engineered to carry certain recombinant properties that cause it to attach and invade human cells very efficiently. And wouldn't you know it; both ivermectin and chloroquine are particularly effective in targeting and disabling RNA-type viruses. Couple this with these medicines general prophylactic (preventative) properties, and the last thing the Cabal wants is the peasantry interfering in THEIR science by self administering alternative cures. Especially when any objective study of how exactly those cures work will point big, fat, accusatory fingers back at those who created and marketed this SARS-CoV-2 bioweapon in the first place.

Another very interesting quote from the Forbes Magazine article says;

> *"Despite the intense scrutiny that coronaviruses have undergone since SARS, it's still not altogether clear why three coronaviruses— SARS-CoV-1, MERS-CoV and SARS-CoV-2 (the source of the COVID-19 pandemic)—have led to far more severe symptoms and a higher mortality rate, while the other four known human coronaviruses remain much milder."* [lxxxvi]

Well, one possible answer is because they've been *designed* that way. After all, the 'discovery' of the four milder coronaviruses spans the period 1965-2005 with a 35-year gap between OC43 in 1969 and NL63 in 2004, with the 'far more severe' SARS-1 and MERS then arriving on the scene between 2003 and 2012, placing the two distinct groupings in an overlapping historical sequence that accommodates the rise in coronavirus research from the 1990's; the artificial creation of SARS-1 in 2002-03; and the commercial rush to unlawfully patent all things coronavirus-related especially after 2003. But maybe SARS-1 and MERS were a little bit overkill for the New World Order because they wouldn't have differentiated between the peasantry and the elites if they got out of control – right? So, those two had to be dropped from consideration for the NWO agenda, and the research discontinued. The four milder coronaviruses only gave us the sniffles, so they weren't going to do the job either. But SARS-CoV-2? Well, just like Goldilocks said, "This one is just right!" At last they had found the highly-infectious but not-really-dangerous pathogen that would not only prompt a global panic, but that could also be used – directly or indirectly – as a vehicle to inject whatever necessary 'stuff' into the population so as to advance The Great

Reset. And here we may have found a slightly more credible explanation for why proper research into the four milder coronaviruses (common colds) has never properly progressed, and why even though SARS-CoV-1 and MERS-CoV were considered 'High Consequence' diseases, that the research into them has also since abated and the pathogens themselves seem to have all-but disappeared from consideration. As one virologist puts it; *"Zoonotic viruses are like that. They do unpredictable things. They tend to appear and disappear at will, and we don't really know why."* Yes, we can all see that now – but unpredictable - really?

Let's have a look at what the experts tell us about zoonotic viruses in particular. Professor Richard Webby PhD., of the WHO explains;

> *"Part of the reason it's hard to predict which viruses will become dangers to humans is because a virus tends to optimize itself for its current host. A virus's main purpose in life is to grow in that host. [Avian flu] (for example) is going to optimize itself to grow in chickens. It's becoming clearer that that optimization for growth in chickens doesn't increase risk in humans. Thus, while it's important to monitor avian flu, a given strain of bird flu most likely won't become a threat to humans. Unless it does."* [lxxxvii]

So, if viruses "optimising themselves" is NOT in fact a threat to humans, then what _is_ increasing the risk for humans then, if this optimisation is not happening 'naturally'? Could the scientists themselves be giving us the answer when they further noted in 2018 that, *"..the danger of zoonotic disease has increased in the past 2 to 3 decades as a number of zoonotic viruses have matured to a dangerous point. (But) Why now?"* [lxxxviii]

Yes, why now indeed? Especially when – as we are about to see in the next chapter – that patents into coronavirus research also coincidentally started to get filed, "2 to 3 decades ago".

Of the various suggestions that are given to account for the surprise and unexpected leap of animal viruses into humans, very curiously, no-one seems to have considered the possibility that all of this messing with animal viruses – including the unlawful gain-of-function research, could be the culprit here? Because the plain fact is that the scientists can't otherwise explain the zoonotic phenomenon – or definitively account for its origins can they? It's all suppositions, and tentative deductions, and guesstimates, but without any solid, empirical, scientific *proofs* of their hypotheses. On the other hand however, the one thing that is eminently predictable is that every time a new virus, or pathogen or pandemic is declared, that more and more money and resources will be poured into those establishments whose very existence is justified by enigmatic and indecipherable phenomena such as the mysterious origins and diagnostics of zoonotic coronaviruses. So, we have motive, we have method, we have opportunity and we have mal-intent. All of the requirements for a criminal spree on gain of function research that reaps millions and billions in profits for the Big Boys.

Then, almost as an afterthought, another proverbial smoking gun gets tossed into the mix. Speaking of, "..the enormous public health threats" that coronaviruses pose to the public, Dr Peter Hotez, Co-Director of the Texas Children's Hospital Centre for Vaccine Development, laments the fact that, "the Big Pharma guys and the biotech companies" showed no interest ten years ago in partnering with him and his colleagues in developing a SARS-coronavirus vaccine.[lxxxix] Really? Why ever not? Especially as Dr Hotez confirms that they even had a vaccine manufactured and ready to go, but that they could not draw funding down from the NIAID (Fauchi) nor from the Gates Foundation because apparently, "no-one was interested in coronavirus vaccines at the time"!? He goes on to explain the risks that people inoculated with SARS-type vaccines could develop something known as "Antibody-Dependent-Enhancement" (ADE) or 'Vaccine Enhancement Disease' which is basically a reaction by the host that was seen in vaccine trials in the 1960's (where some children died) as well as in recent Covid-19 animal trials, whereby the vaccines basically caused more disease and damage than any protection they were providing. (Full explanation here).[xc]

So, even though the big biotech labs were openly acknowledging that, *"..the danger of zoonotic disease has increased in the past 2 to 3 decades as a number of zoonotic viruses have matured to a dangerous point.."* and even though they are all recipients of massive public and private funding into vaccine research, still, Fauchi, Gates, Baric, Zhengli, Daszak & Co., including Big Pharma and the biotech giants it seems, that none of them were even remotely interested in Dr Hotez's coronavirus research nor in the vaccine he had created? Need we say it again? Something very fishy is afoot again, and I suggest that there can only be three possible explanations for this:

(i) That the US Government and the biotech industry has no real interest in investing in research purely for the public's benefit.

(ii) That Dr Hotez is not ensconced in the Cabal, so due to professional rivalries they will not fund him.

(iii) That Dr Hortez's invention competes with Cabal-affiliated profit-making patents and could therefore interfere with certain aspects of the overall preplanned pandemic.

It does seem clear that none of the publicly-funded biotech labs are even remotely interested in genuine scientific research 'in the public interest' unless there is private profit to be made from it. They are not interested in finding cures in other words, but they are very much interested in how they can use or manipulate coronaviruses for commercial or political ends. This would not only account for the arrival (accidental or otherwise) of zoonotic pathogens out of commercial animal vaccine production in the 1960's; but it would also account for the lack of public-spirited research over the last 50 years, as well as explain away the lack of interest by the Big Boys in Dr Hotez's research. Because of course, if any such vaccine was brought to market then he would not only have held the patents, and would therefore reap the profits, but any such not-in-the-club vaccine might actually be authentic and effective at whatever it was supposed to do.

Given Dr Hotez's further evidence before the Texas State Science Committee that he wished he had been funded some ten years ago so as to have been ready for the Covid-19 pandemic, because as everybody knows, you can't just pop up and deliver a safe and effective vaccine in a very short period, but that it must first undergo a lengthy, rigorous period of testing... then we must presume and assume that Dr Hotez was not in on the grand scam, and that he might even have been considering delivery of his vaccine in the general public good? We don't know if this is actually the case, but it seems a reasonable conclusion given that he now appears to be working with Indian labs to get another Covid vaccine approved over there. This would of course be yet another reason for competing US-based interests not to want to fund him.[xci]

But to return to the Forbes Magazine quote above, we see again the same type of pseudo-science being used to bamboozle the public as to the purported 'seriousness' of SARS-CoV-2. What do they mean by SARS-CoV-2's 'higher mortality rate' because percentage-wise it has proven no worse that the seasonal flu – has it? So, that 'higher mortality rate' is what exactly? Higher than which other pathogen? What does 'higher mortality' mean in real, social terms? It is accepted of course that of the seven listed coronaviruses that the SARS and MERS variants cause more fatalities per infection, but remembering that the influenza virus is NOT a coronavirus, but that it caused around half-a-million deaths worldwide annually (before it mysteriously disappeared off the charts in 2020) but it still didn't qualify as 'a pandemic' whilst some of these purportedly 'novel' coronaviruses with only a mere smattering of deaths, do?

Or malaria causing an average 400,000 deaths annually, most of whom are children under 5 yet still being 'only' an endemic disease? Well, we simply have to ask why Forbes Magazine is connecting a purported 'higher mortality rate' with Covid-19 when it doesn't even match the normal morbidity rates of the seasonal flu, and certainly has NOT resulted in any overall average annual increase in respiratory-infection-related deaths compared to the morbidity figures of the last decade? And this, notwithstanding the controversy over PCR Testing and the manner in which 'cases' are being recorded, and with the question still hanging in the air as to what exactly Covid-19 is if it *isn't* a naturally-occurring zoonotic virus mutation that has somehow displaced (or in some way has incorporated?) the normal cold-and-flu season, which as we said, appears to have magically disappeared – or has retired from active duty – upon the explicit instructions of the World Health Organisation and Big Pharma that this is now 'Covid time' – and all of the other common viruses need to join right in like good little soldiers, or to get out of the way!

How else, other than by foreknowledge and foreplanning for instance, do we account for 'Product No.902780' being listed on the World Integrated Trade Solution Database (WITF) as, *"COVID-19 Diagnostic Test instruments and apparatus"* that had apparently, somehow been traded internationally since 2017 – a full 2 years before 'Covid-19' even existed!? ...and which designation (as screenshotted below) has since been changed, in the wake of public alarm and frantic fact-checker efforts from all of the usual suspects including Reuters / FactCheck / FullFact & Co., *et al*, to the far more innocuous-sounding and far less incriminating designation: *"Instruments used in clinical laboratories for In Vitro Diagnosis"*?

Reporter	TradeFlow	ProductCode	Product Description	Year
European Union	Export	902780	COVID-19 Diagnostic Test instruments and apparatus	2017
United States	Export	902780	COVID-19 Diagnostic Test instruments and apparatus	2017
Germany	Export	902780	COVID-19 Diagnostic Test instruments and apparatus	2017
Japan	Export	902780	COVID-19 Diagnostic Test instruments and apparatus	2017
China	Export	902780	COVID-19 Diagnostic Test instruments and apparatus	2017
Hong Kong, China	Export	902780	COVID-19 Diagnostic Test	2017

WITS
World Integrated Trade Solution

229

PATHOGENIC PATENTS

A SIMPLE CASE OF A, B & C

We return now to the issue of patents and how this undeniable, documented evidence 'on the historical record' will help us piece together the whole Covid-19 phenomenon narrative – from all differing sides of the debate – into an understandable explanation of what's really going on – or at least, in a form that will help us to eliminate what's NOT going on; despite what 'they' keep telling us.

The Damning Patents. It's time to explain what SARS-CoV-2 actually is, because as we are about to see, it is neither 'novel' (new) nor natural. Although it probably doesn't have to be repeated at this stage, I feel it's appropriate to remind everyone that we should NOT be delegating our discernment to others – not even to the contents of this book – but should take ownership of our own minds and spirits and weigh up all of the combined facts and evidence against all of the desperate fact-checking and debunking being so expertly unloaded upon us by professional liars and obfuscators.

The whole notion of The Triadic Archetype as laid out in Part One is to get us all thinking about an alternative reality; one where we do not *need* authorities, and experts and governments telling us what to do, because we ourselves are possessed of a singular personal authority, well-informed and intuitive that resonates with truth, knowledge and wisdom, rooted in compassion and kindness and generosity, and all of those other noble humane traits that separate us from 'the beasts'. This may be a radical departure from tradition, but as-and-when we discover that 'tradition' (whatever form it may take) is anathema to goodness and truth, then we have an obligation as true human beings to have the courage and resolve to move away from those traditions and to try to conceive of something better.

Anyway, to set the stage for our patents exposé, we first need to remember that it was commercial incentives that first prompted scientists at the NIH and the NIAID (in the USA) as well as in UK labs, to start studying animal coronaviruses back in the 1930's. We then have a very questionable and somewhat ambiguous provenance as to how human coronaviruses actually emerged / were discovered / mutated into existence etc., etc. But the main point to note is that the scientific establishment has been aware of the existence of human coronaviruses for quite some time, and as we have already explained, a handful of highly-specialised experts have been tinkering with them now for well over 20 years. One of the chief tinkerers was Dr Anthony Fauchi who had been working on AIDS-related projects for some time when he discovered that the human coronavirus was a good vector for introducing 'stuff' into human cells. 'Vector' in this case means an injectible vehicle that has certain abilities to by-pass natural defences and such like, so as to be able to deliver 'stuff' into our cells.

Now I'm not even going to try to get into any AIDS-related discussion here because first of all I am obviously no expert in the field and the AIDS debate has been laced with all of its own controversial theories down the years about what AIDS really is; about who or what caused it; and whether or not it emerged naturally or was engineered, etc.[xcii] For our purposes today we need only note that Dr Fauchi is considered a world expert on AIDS and so it should be no great surprise to discover that he was interested in how coronaviruses could be used in a HIV vaccine – possibly as a prophylactic (a preventable against disease). On the assumption that this was noble, well-intentioned work, we note the following succinct quote that gives us laypeople some insight into how the Human Immunodeficiency Virus (HIV) causes Acquired Immunodeficiency Syndrome (AIDS):

> "HIV particles do not cause AIDS, our own immune cells do: The virus turns host immune cells into suicide machines, using them to spread the virus and cause the progression from HIV to AIDS."
>
> Science Daily. August 2015.

So, a virus can cause our own healthy cells to turn into 'suicide machines' can it? An interesting reality that was known by the experts which we should file away for later reference.

The next thing we have to understand is how scientists 'sequence' or chart the essence of any particular organism. Now we have all seen pictures of human DNA in the twisty double-helix shape, which, if we were to untwist and straighten it out would look like a ladder, or better still a set of railway tracks, because after all, we are talking about super-long strands of DNA with multiple thousands of base-pairs, such as the steps on a ladder or the sleepers in a train track. So, the DNA (and/or RNA) of coronaviruses is usually around 30,000 base pairs, and it is by looking at the complete sequence that we get to identify 100% what it is exactly that we are looking at. So, with apologies for the over-simplistic explanations, if we imagine six sets of train tracks lying parallel to each other, we would be able to identify each of the six (pre-Covid-19) coronaviruses by the sleeper patterns, or by the colours, qualities or textures (for example) of their tracks. A 'DNA signature' if you like that tells us which virus is which.

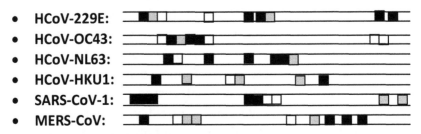

- HCoV-229E:
- HCoV-OC43:
- HCoV-NL63:
- HCoV-HKU1:
- SARS-CoV-1:
- MERS-CoV:

Diagram is purely for explanation purposes

Now this is where things get interesting, because even though one is not supposed to be able to *lawfully* patent nature, somehow the following SARS-related patents have in fact been filed, meaning either they are NOT 'natural' occurrences, or, that some of the most eminent scientists and academics in the world have been on a thoroughly-illegal patent-registering spree, just because they felt like it? Again, we credit Dr David Martin of M.Cam Inc., for his tireless research and courageous publication of many of these facts. Not being molecular scientists, it is not realistic nor practical for us to try to explain the full contents of these patents, but we urge everyone to check them out via the end-noted links provided. Because the usual troop of hired fact-checkers have already been hard at work trying to undermine the documented facts and the logical conclusions that they point to, by either targeting those who have spoken out or, by fogging and obfuscating the conversation with even more complex counter-arguments. But again, we ask the reader to simply consider what is being presented here against the backdrop of the Great Lie and allow one's moral intuition to simply 'do the math'.

The 'priority date' or date of first registration of each patent application is listed in [square brackets]. This date is actually more important than the filing date for the purposes of our research, because it proves knowledge of the patent's contents at least from that priority date. Abstracts are included for the more technical minded, but don't worry if that's just too much detail because it's the titles of the patents and the priority dates that are most pertinent.

(i) Patent No US-6372224-B1. [14th Nov 1990] Filed January 28th 2000 by Pfizer et al. *"Canine coronavirus S gene and uses therefor."*

Abstract: The present invention provides the amino acid and nucleotide sequences of a CCV spike gene, and compositions containing one or more fragments of the spike gene and encoded polypeptide for prophylaxis, diagnostic purposes and treatment of CCV infections. [xciii]

(ii) Patent No US-6593111-B2. [21st May 2000] Filed 21st May 2001 by Ralph Baric & University of North Carolina (UNC). *"Directional assembly of large viral genomes and chromosomes."* [xciv]

Abstract: Full-length, functionally intact genomes or chromosomes are directionally assembled with partial cDNA or DNA subclones of a genome. This approach facilitates the reconstruction of genomes and chromosomes in vitro for reintroduction into a living host, and allows the selected mutagenesis and genetic manipulation of sequences in vitro prior to reassembly into a full length genome molecule for reintroduction into the same or different host. This approach also provides an alternative to recombination-mediated techniques to manipulate the genomes of higher plants and animals as well as bacteria and viruses.

(iii) Patent No US-7279327-B2. [20th April 2001] Filed 19th April 2002 by Ralph Baric & UNC. *"Methods for producing recombinant coronavirus."*

Abstract: *A helper cell for producing an infectious, replication defective, coronavirus (or more generally nidovirus) particle cell comprises (a) a nidovirus permissive cell; (b) a nidovirus replicon RNA comprising the nidovirus packaging signal and a heterologous RNA sequence, wherein the replicon RNA further lacks a sequence encoding at least one nidovirus structural protein; and (c) at least one separate helper RNA encoding the at least one structural protein absent from the replicon RNA, the helper RNA(s) lacking the nidovirus packaging signal. The combined expression of the replicon RNA and the helper RNA in the nidovirus permissive cell produces an assembled nidovirus particle which comprises the heterologous RNA sequence, is able to infect a cell, and is unable to complete viral replication in the absence of the helper RNA due to the absence of the structural protein coding sequence in the packaged replicon. Compositions for use in making such helper cells, along with viral particles produced from such cells, compositions of such viral particles, and methods of making and using such viral particles, are also disclosed.[xcv]*

(iv) Patent No US-46592703-P. (Registration Number still exists on the database affiliated with patents (iv) & (v) below, but otherwise, does not return any search results).

(v) Patent No US-7220852-B1. [25th April 2003] Filed 12th April 2004 by CDC & US Government. *"Coronavirus isolated from humans."*

Abstract: Disclosed herein is a newly isolated human coronavirus (SARS-CoV), the causative agent of severe acute respiratory syndrome (SARS). Also provided are the nucleic acid sequence of the SARS-CoV genome and the amino acid sequences of the SARS-CoV open reading frames, as well as methods of using these molecules to detect a SARS-CoV and detect infections therewith. Immune stimulatory compositions are also provided, along with methods of their use.[xcvi]

(vi) Patent No US-7776521-B1. [25th April 2003] Filed 14th May 2007 by CDC. *"Coronavirus isolated from humans."*

Abstract: [Identical wording to the abstract above].[xcvii]

(vii) Patent No US-7151163-B2. [28th April 2003] Filed 28th April 2004 by Sequoia Pharmaceuticals Inc. *"Antiviral agents for the treatment, control and prevention of infections by coronaviruses."*

Abstract: The invention provides compositions and methods that are useful for preventing and treating a coronavirus infection in a subject. More specifically, the invention provides peptides and conjugates and pharmaceutical compositions containing those peptides and conjugates that block fusion of a coronavirus, such as the SARS virus, to a target cell. This blocking mechanism prevents or treats a coronavirus infection, such as a SARS infection, in a subject, such as a human subject.[xcviii]

(viii) Patent No US-7618802-B2. [21st July 2003] Filed 19th Jan 2006 by Ralph Baric et al, & UNC. *"Compositions of coronaviruses with a recombination-resistant genome."*

Abstract: The present invention provides a cDNA of a severe acute respiratory syndrome (SARS) coronavirus, recombinant SARS coronavirus vectors, and SARS coronavirus replicon particles. Also provided are methods of making the compositions of this invention and methods of using the compositions as immunogens and/or vaccines and/or to express heterologous nucleic acids.[xcix]

So what does all this mean, exactly? Well, in simple terms it means that certain core individuals and agencies were unlawfully registering patents on illegal bioweapons-type coronavirus research that has been ongoing since at least 1990 [coincidentally matching the 2 or 3 decades where zoonotic viruses are reportedly on the rise] and – we must logically surmise – that these agencies and individuals were then licensing other individuals, corporations and government bodies worldwide to use their methods in similar 'gain of function' research (for profit) such as the 100+ entries on this linked M.Cam document which drew its results from two pieces of Covid-19 related scientific literature: [c]

(i)   *"A novel bat coronavirus reveals natural insertions at the S1/S2 2 cleavage site of the Spike protein and a possible recombinant 3 origin of HCoV-19."* [ci]

(ii)   (and) *"The Proximal Origin of SARS-CoV-2."* [cii]

This not only means that the Covid-19 phenomenon is the direct by-product of prohibited bioweapons-type activity, but also that the patent holders had 100% control over all things coronavirus-related for a number of years well in advance of SARS-CoV-2.[ciii] When we remember that SARS-CoV-2 is only called that because of its alleged similarities with SARS-CoV-1, and we then read that Dr Ralph Baric from UNC had used his funding from the NIH to synthetically alter coronaviruses for the express purpose of general research, for pathogenic *enhancement* (gain-of-function), and for potential therapeutic interventions (vaccines and cures) eventually coming up with a *composite* of assorted viral fragments which we now know as SARS-CoV (No.1) where he explained in a 2003 paper:

"Using a panel of contiguous cDNAs that span the entire genome, we have assembled a full-length cDNA of the SARS-CoV Urbani strain, and have rescued molecularly cloned SARS viruses (infectious clone SARS-CoV) that contained the expected marker mutations inserted into the component clones." [civ]

In other words, that Ralph Baric and friends – amongst whom we must surely include Anthony Fauchi, Peter Daszik, Shi Zhengli and their generous donors at the Gates Foundation all knew that a composite / cloned / manufactured / bastardised version of SARS-CoV existed for nearly 20 years, at very least. Not forgetting either that Dr Ralph Baric was also on the World Health Organization's International Committee on the Taxonomy of Viruses (ICTV) and a member of the Coronaviridae Study Group (CSG), in which privileged positions he was able to influence decisions that directly benefited himself, his colleagues and the various agencies and corporations they were employed by; including the CDC, NIAID, WHO, UNC, Johnson & Johnson, Sanofi, Moderna, Ridgeback, Gilead, and Sherlock Bio-sciences for example, thus creating a sort of undeclared biotech conglomerate-monopoly that ensured that the millions in Federal grants and other fundings from the likes of the Gates and Clinton Foundations would continue to flow in all the right directions. Adding even more intrigue to the story is the reported fact that in 2007 [Patent No (v) or (vi)] the CDC attempted to patent the same viral sequence as the Pfizer patent of 2000 [No (i)] but their application was refused on the grounds of repetition. Then, the CDC paid the Patents Office through some highly questionable privatisation process to file their patent applications away from the public view. Now why would they do this if, as the fact-checkers tell us, the CDC only registered these patents in order to protect the research "in the public interest"? Perhaps Dr Julie Gerberding, Executive Vice-President of Merck, who was the Director of the CDC at the time, could answer that one for us? Because according to Karl Gustav Jung, when coincidences keep happening, they are no longer coincidences.

Writing to the Governor of Montana in a formal registered complaint dated July 2021, Mr David W Rowell says: *"Since 2003, the U.S. Department of Health and Human Services and their subordinate organizations - the National Institute of Allergy and Infectious Diseases (NIAID) and the Centers for Disease Control and Prevention (CDC) - maintained a patent preventing any independent organization from testing for the presence of coronavirus transmissible to humans through 2018 resulting in a complete lack of testing technologies;"* [cv]

In line with our own research, Mr Rowell goes on to list as collaborators; "..the CDC; NIAID; University of North Carolina Chapel Hill; Wuhan Institute of Virology; National Institutes of Health; U.S. Department of Health and Human Services; the President's Task Force and all US State Governors except those of North Dakota, Nebraska, Arkansas, Utah, Wyoming, South Dakota, and Oklahoma."

Now to be clear, it appears that these various patent filings only definitively and empirically 'prove' the following things:

- That all of the discrete elements and components of SARS-CoV-2 were known well in advance by certain key players.
- That the S1 spike protein in particular (which is supposed to be novel) had been patented by Pfizer in 1990.
- That research into 'defective, recombinant coronaviruses' was ongoing by Ralph Baric *before* the SARS-CoV-1 outbreak.
- That in 2003 Dr Baric registered a patent on a *composite* (made in the lab) full-length version of SARS CoV-1.
- That all of the processes and procedures necessary to create SARS-CoV-2 (if someone so wished) had therefore been documented and patented long before the 2020 pandemic.
- That the research itself and the patents lodged were variously and technically unlawful for the reasons already explained.
- That anyone wanting to conduct coronavirus-related research would need a licence to do so from the patent holders.
- That the CDC sought to privatise their patent applications and shield them from public access – and then lied about it.
- That incredibly, that Sequoia Pharmaceuticals [Patent No (vii)] somehow managed to produce a remedy for the SARS-CoV virus just a few days after the CDC patented SARS-CoV-1 [Pat. No (v)].

That particular Patent (No (vii) is worth a full read, because it not only explains how Sequoia had come up with their cure based upon the composite sequences of Dr Baric and the UNC, but, making reference to the other usual suspects, it also definitively states:

> "Scientists at the Centers for Disease Control and Prevention (CDC) and other laboratories around the world have detected a previously unrecognized coronavirus in patients with SARS. The evidence for a coronavirus was based on genetic fingerprint and electron microscopic ultrastructural studies and was widely reported in the popular press. Virologists at the CDC, WHO and numerous academic laboratories all reported that a coronavirus is the leading hypothesis for the cause of SARS."

"Studies? Reported on by the popular press? An hypothesis??" And this is what they based their miraculously-swift research and SARS cure upon? Sounds very, very unscientific to me. But perhaps it should also be noted that Sequoia Pharmaceuticals was only founded in 2002 (just before the SARS No 1 outbreak) and has since (reportedly) been absorbed into the holdings of Pfizer and Janssen.

In short, and in context of all of the information presented so far, including the continued absence of definitive, empirical proof from a trustworthy source that SARS-CoV-2 originated in an animal before being transmitted 'naturally' to us; this patented research establishes beyond doubt that SARS-CoV-2 was neither 'novel' nor 'natural'. In Patent No (iii) for example, funded by Dr Fauchi's NIAID and first lodged in 2001, Dr Ralph Baric of UNC further clarifies that they were already working on; "An infectious replication defective coronavirus (which is) designed to target human lung epithelium." A disease-causing, cloned and bastardised virus that would attack human lungs in other words. They had already worked on spike proteins and had already worked out how to use computer code to turn 'scourge' organisms (such as these now-defective coronaviruses) into disease-causing pathogens – so apart from there being nothing truly 'novel' about the arrival of SARS-CoV-2, Dr Baric's patent application leaves us in no doubt as to what the intentions for this research were; to deliberately cause respiratorial sickness and disease in human hosts.

Further underscoring the 'not natural' elements; the very fact that the spike proteins in SARS-CoV-2 (forms of which were being researched by Pfizer as far back as 1990) are more efficient at binding to human cells than they are to any other species, forces the common-sense conclusion that in order to be *most* effective at attaching to human cells vs animal cells, that SARS-CoV-2 either had to *originate* in humans in its already highly-developed and perfected state – something that isn't even being remotely postulated by the biotech-scientific establishment – or, that it was artificially engineered by some clever and devious species with more laboratory know-how than your average bat or pangolin.

If we return briefly to our railway analogy, it appears clear that SARS-CoV-2 was 'created' by taking some existing sections from the known human-infecting coronavirus sequences and then laying them out if-you-like as a seventh composite railway line – but one that had several massive gaps in it, and then getting a computer to generate the remaining code necessary so that SARS-CoV-2 would first become (a) a theoretical reality; (b) then a computer-generated-and-coded reality that could then be definitively referred-to in scientific papers and media releases; whereupon (c) the CDC/UNC/WIV boffins could then actually create a tangible sample, which, in the absence of any genuine isolates, they could then culture and dispatch these newly-fabricated SARS-CoV-2 samples to laboratories all over the world. Meanwhile, the WHO was advising everyone to 'test, test and test' for something which *could* now be scientifically *described*, but which didn't actually need to be present in any human hosts for the PCR Tests to be triggered, because as we now know, those PCR Tests were going to pick-up on *any* coronavirus or seasonal flu debris as we explained in the analogy of the mice in the field. In other words, the PCR Tests being advocated and pushed by the WHO didn't even need the presence of SARS-CoV-2 in the host to register 'a Covid-19 case'.

Existing virus strands:

- **HCoV-229E:**
- **HCoV-OC43:**
- **HCoV-NL63:**
- **HCoV-HKU1:**
- **SARS-CoV-1:**
- **MERS-CoV:**

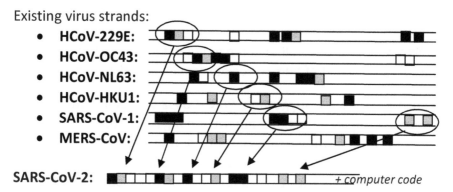

SARS-CoV-2: _+ computer code_

Diagram purely for explanation purposes.

Obviously, if SARS-CoV-2 can be created like this – as a composite of other viral strands that have been stitched together with some computer code and then cultured in the lab – just as the SARS-CoV-1 composite was in patent No (iii) – then so can any possible 'new variants' – right? With the Gates-funded WHO and the CDC calling the shots (excuse the pun) all they have to do is get some compliant agency, lab or government department to announce 'a new variant' in their locale which can then be mapped and sequenced in the lab without there ever needing to be an actual *physical* sample of any such variant on the ground. On the other hand, if they DO need to produce a physical mutation for the questioners and the naysayers, well, they already have the know-how and in any event, because SARS-CoV-2 isn't dangerous anyway, it's no big deal to have to create this-that-or-the-other variant or mutation, if that's what's needed to keep the momentum going. Thus, we get Alpha, Beta, Gamma & Delta variants leading the charge but only as "variants of concern" with Eta, Iota & Kappa as "variants of interest" lining up neatly behind them with a couple of other as-yet unnamed mutations, ready to be elevated to "variants of high consequence" should we need another injection of momentum (excuse the pun) into the Covid-19 narrative. The recently-aired suggestion by an increasing number of concerned medical professionals that these 'variants' are in fact being fuelled, driven or even directly caused by the Covid-19 vaccine rollout, is gaining traction and obviously should NOT be ignored, especially when any such scenario fits so neatly

241

with an agenda that is clearly about getting the whole of the global population injected, passported and duly 'marked-and-registered' on the New World Order's all-encompassing database, so that the not-so-great Great Reset of human society remains on diabolical track.

Because if, as we are suggesting, SARS-CoV-2 was deliberately man-made to be highly infectious but NOT dangerous, then this would also fit with a narrative that has the authorities making general, comparative reference to 'SARS-CoV' at the outset of the crisis so as to instil fear and alarm in the public to get the diabolical plan up-and-running at a sprint. They only had to mention the words 'dangerous virus' to trigger off our natural, yet carefully-cultivated subliminal fears of impending doom. After all, with-or-without SARS-CoV-2 people do tend to die in large numbers every single year. With the PCR Tests picking up on all of the normal cold and flu cases and thereby providing sufficient numbers to establish 'evidence' of a purported pandemic, the only real question we need to be asking now is whether or not it was a genuine release (accidental or not) of SARS-CoV-2 itself that started the panic in Wuhan and which then went on to spread naturally around the globe, or, whether there could be some other explanation for the pathogen release that resulted in the pandemic-priming events in Wuhan?

Again, there are a couple of common-sense issues to bear in mind here; the first of which is the obvious fact that after they saw the potential of a viral release to launch their 'Great Reset' that the diabolicals who are engineering and driving the pandemic narrative realised that they too would be at personal mortal risk if a genuinely dangerous coronavirus was released into the air, such as SARS-1 in 2003 and then MERS in 2012 with mortality rates of around 12% and 34% respectively. So, whatever SARS-CoV-2 was going to be, it wasn't going to be a *genuinely* dangerous pathogen – at least no more so than a common cold or the flu – as has since proven to be the case. As we already said, with PCR Tests picking up huge numbers of false-positives, all they had to do was present a *theoretical* version of SARS-CoV-2 and marry it to the normal cold and flu season to be able to convince the population that we were all doomed, and that 'thousands of people' were 'dying from Covid-19'. This not only explains the absence of an *original* isolate (as per Koch's Postulates) and no originating animal host yet being found (nor will it); but it makes perfect sense to me in a world where the Cabal does indeed control all of the important major institutions in society, most notably the news media.

You see, whatever happened, the Cabal had to have either huge numbers of reported 'cases' and/or an undeniable concentration of localised infections complete with shocking, sudden deaths to get things going. Because no-one was going to panic and submit to tests and lockdowns if the new mystery disease was just another relatively harmless strain of the common cold or flu or such like – as it has since turned out to be. So, in case the ridiculous PCR Test didn't provide enough numbers for them to convince us that everyone would soon be dropping like flies, well, they could then simply inform us that believe it or not, we could be completely asymptomatic (showing no signs of sickness) yet *still* have the disease. Really? Yup! And they also had a plan if the population of the world didn't jump when they barked the first time and if, after a year-or-so, they started to question how genuinely 'dangerous' the virus was, or refused to get injected; because that would be the cue for some mysterious and ever-more-dangerous variants to appear, complete with punitive restrictions on those who hesitated to take the tests, the injectibles and the variant-specific boosters, with the media still fuelling the fear and alarm...

One wonders what level of credulity one must possess to keep on swallowing this guff!

*   *   *

Anyway, so now we have a credible, if highly disturbing formula as to how and why a highly-infectious but generally non-lethal virus could have been deliberately released in Wuhan so as to trigger off a flawed and baseless pandemic. This would mean of course that

SARS-CoV-2 is now at large in the population and is causing *some* cold-and-flu like disease – but surely, it's not nearly sufficient disease to shut down society, is it? No, obviously not. But as long as it's sufficient to carry the narrative, then that'll do fine – right? On the other hand, we also now have a formula whereby SARS-CoV-2 possibly *only* exists in actuality in those laboratories where the CDC sent their manufactured-cloned composite samples – something they claim to have done in January 2020 – in spite of their admitted absence of any proper isolates at the time!? This would mean that SARS-CoV-2 is NOT in fact at large in the population, but only exists in theory and in the reconstructed samples that are now circulating amongst the research labs, providing all of the necessary detail for tame scientists and a well trained media to debate, and for the rest of us to be utterly confused by; as the pandemic juggernaut gathers steam and momentum. Meanwhile, all of the injuries, deaths and variants that are being caused by the injectibles get reported not as they are – as the consequences of getting 'vaccinated' with bio-weapons – but instead as supposed 'new Covid cases', well, we can see how the diabolical cycle is now well underway. But if this latter scenario is the case, then we still have to account for the original panic out of Wuhan and for the subsequent lockdowns of a number of major cities at the start of the pandemic, because if it wasn't physical SARS-CoV-2 that was causing it, then what was it?

The recent publication of a US Congressional Report may hold some clues, although the very fact that it is coming from 'official sources' must be kept in mind. Basically, the Report alleges an international cover-up of the true origins of the virus by the Chinese Communist Party that is supported by 'suspicious activity' at the Wuhan Institute of Virology (WIV) in September 2019, and by the active efforts of the CCP since to thwart any efforts by the international community to get at the facts. But the truth is that the Covid-19 agenda (whatever it may ultimately prove to be) is being well-served by such obfuscatory allegations and dare we say it, by any officially-sponsored conspiracy theories, and we should remain alert to that possibility too.

Now, I am quite aware of course that this book will likely come in for some close attention from the fact-checkers and debunkers, and I am loathe to provide them with anything that might undermine the overall message here, which is essentially about the battle between good and evil and of our own individual roles in that conflict. But another theory has arisen recently that would fit in with the truly diabolical character of the worst elements of the Cabal. The possibility that it wasn't SARS-CoV-2 at all that was released in Wuhan (because it was possibly still only a theoretical construct at the time) but that some other pathogen – more individually dangerous than any pesky lab-created coronavirus – was selectively released for the express purposes of triggering off the pandemic narrative. That trigger-pathogen may have been viral too, only with a limited shelf-life and contagion-reach so that only a limited number of people could, or would, be affected. And if it worked in Wuhan, then it could be deployed elsewhere in key population centres such as Milan, Italy; Seattle Washington; and Isfahan, Iran, so as to get everyone's undivided attention. But again, as long as those numbers were sufficient to get the ball rolling, well, they had spent an awful lot of time, money and effort planning for all this, so who really cared how it started – as long as it started on schedule. This theory is just speculation at the moment, but as we've said now many times, when you're dealing with profit-driven sociopaths, anything is possible.

Perhaps the whole sorry, 'Let's get everyone vaccinated' exercise can be summed up by the internal statement by Dr Peter Daszik in 2015:

"We need to increase public understanding for the need for medical countermeasures such as a pan coronavirus vaccine. A key driver is the media, and the economics will follow the hype. We need to use that hype to drive profits. Investors will respond if they see profit at the end of the process."

Yes Friends. It's always been about the moolah.

CHAPTER NINETEEN

BEASTS IN BUSINESS SUITS

THE GREAT RESET & THE NEW WORLD ORDER

Throughout the book we have intermittently made reference to The New World Order, the World Economic Forum, the Fourth Industrial Revolution and The Great Reset in context of the Covid-19 phenomenon, and it is important for all of us to see how these futuristic initiatives by these self-appointed elites tie in to the current crisis, not so much in their declared humanistic ideologies and the projected 'benefits' for humankind, but more realistically, in terms of the effects of those troubling initiatives on the hearts, minds and souls of all of us – even of the so-called 'diabolicals' themselves who are driving the agenda.

I have been very fortunate in life. I grew up in different locations in Ireland and the UK where most of my free time was spent outdoors playing football or just roaming the nearby hills and fields – often alone – as I explored the hedgerows and copses for whatever I might discover. This exploratory bent continued in my adult years where if I wasn't driving trucks into the USSR or preparing expeditions to Nepal, I was working in the Alps or canoeing down ravines, or camping in the desert en-route to some exotic destination. It was all 'work' of course, because there was no other way to satisfy my yearning for adventure and travel other than by working for tour operators and transport companies because otherwise, I could never have afforded it. The point I am making is that I have a solid and profound respect for nature and for all things natural, and for all of the colourful aspects of diverse cultures, and I remain deeply concerned at the way the world is heading – not so much in respect of this Covid-19 phenomenon which of course is THE issue of the day – but in respect of the last 100 years or so, where it is clear that we simply cannot continue living this way and still expect nature to continue putting up with us. By this, I mean that in respect of our stewardship of the planet that mankind is acting in a manner that is akin to the medical dictionary's various definitions of cancer as: *"A malignant growth or collection of abnormal cells resulting from uncontrolled reproductions that are hard to contain or eradicate, and which infect and consume their immediate environment."* Also referred to in the standard dictionary as, *"An evil or destructive practice."*

Harsh words indeed to describe humanity, but arguably not inaccurate inasmuch as we are undoubtedly raping, pillaging and plundering the earth and its natural resources for immediate, short-term monetary again and are destroying this beautiful world in the process. By 'we' of course, I mean chiefly the Cabal, and all of us other ignorant, apathetic or lazy souls who are NOT speaking up or fighting back, or who feel comfortable enough just doing as we're told and going along for the ride, as long as we feel that our individual lives are not being directly affected – or so we naively think and believe.

This brings us back to our discussion in Part One about taking personal ownership of our own lives in so many ways, rather than delegating responsibility away to those who would abuse our trust to do terrible things 'in our names' – as long as we keep allowing them to do so, and as long as we keep funding the operation.

The point I am making here is there is no doubt we are making a hash of things right now and that society could be organised and managed in many more productive and harmonious ways that would not necessarily result in wholesale pollution, in the depreciation of wildlife, in the poisoning of the oceans and the ravaging of the rainforests; that there are better methods of collective stewardship that will solve the problem of world poverty, the unequal distribution of wealth and resources, the cessation of wars and a genuine application of justice. All of this I am in favour of – for obvious reasons – but as with any problem, or ailment, or disease, or destructive practice, when it comes to providing for a solution, that solution must NOT be worse than the problem itself. This immediately brings us into the realm of objectivity and subjectivity and the manner in which we decide what is 'right' and what is 'wrong' – or indeed what may be 'better' or 'worse' solutions to any given problem, because if you recall our discussion in Part One the central theme was that there are in effect, two sets of people who are fundamentally, diametrically and philosophically opposed as is shown in the diagrams on p.62 & p.94.

On the one hand we have intelligent, well-informed and driven empaths, and on the other side we have the sociopaths and psychopaths – equally intelligent and driven – along with all of the lackeys and toadies that serve them, doing their utmost to literally take over the world. And make no mistake, they will achieve this by any means at their disposal regardless of what we may believe is morally right-or-wrong, including using lies, tricks and deceptions without a second thought. This longstanding strategy comes into play with the Covid-19 phenomenon inasmuch as; (a) the Cabal is *claiming* that all of the official reactions and the impositions on our civil liberties are of course necessary for 'the greater good'; and (b) that the consequent move towards totalitarianism and establishment-controlled fascism by way of the New World Order is likewise being sold to us on the premise that this is the best and *only* way forwards to save the planet and ensure the fruitful advancement of human society. But just in case we haven't made it clear enough already, these declarations and pronouncements are no more than the sophisticated lies of a gang of interconnected sociopaths whose eyes have never strayed from the prize of total ownership of everything and everybody on the planet. If that ownership requires a population cull, then so be it. If it requires the abandonment and disposal of pretend-democracies and republics, well they can go too. And if that ownership requires that every person on the planet needs to be tracked, traced and vaccinated, well, you can finish that sentence for yourself... But as is shown in the diagrams on p.62 & p.94, in order to achieve this they will need the support and complicity of the apathetic masses – either through their ignorance, their conditioning or their compliance, or, in fear of what will happen if they do not follow the program. Or contrarily; in anticipation of the expected rewards, privileges and benefits when they do – such as we see happening in Ireland already where the official narrative went from a simple need to "flatten the curve" for two weeks, to the introduction of Covid passports and 'vaccine bonuses' and other privileges for those who obey. These are Étienne de La Boétie's knaves and dupes who have, between them, kept lineages of tyrants, dictators and despots in power throughout history, and it's well past time for the rest of us apaths to stand up and begin to do something about it!

In other words, whatever the collective problems we are facing there will be 'better' and 'worse' ways of solving them. But we can't rely on the moral values of the sociopaths and psychopaths to make those crucial discernments because they don't have active morals and will decide what is 'better' or what is 'worse' according to their own perverse ambitions. Don't forget, that these people see the rest of us merely as products, or as disposables. No, the key to deciding what is 'better-or-worse' in any given situation lies in the application of *humane* values, and we need intelligent, driven empaths to be doing that because at the very top of that list is the sanctity of human life. In other words, in the understanding that we all have the capacity to embody the Divine, then we must acknowledge and respect the potential of each and every person alive to carry and embody universal truth in an intelligent, self-conscious, humane level which is not possible for any other aspect of creation that we know of. Indeed, looking again at The Triadic Archetype we see that it was those people who were pure of spirit with well-informed minds who drove the quest for truth, justice and equality down through the ages. It was the prophets, the messiahs, the great philosophical teachers and pure spiritual thinkers as opposed to money-grabbing politicals and tyrannical despots who embodied the nature of true humanity; and it is that intrinsic yearning in each of us – or at least in those of us who have not yet had their consciences conditioned, bribed or bullied out of them – to aspire to truth and justice and decency; that urges us each to act, and to object and to resist, whenever we see gross injustices or witness the abandonment of truth and of humane values.

If the burgeoning population of the world is 'the problem' for example, then let's solve that problem by humane means. If pollution, or war, or poverty are the problems, then let's solve those problems too by humane means; and if microscopic diseases are the problem, let's tackle them as well on a humane basis. But we're not going to do it by handing over our authority to the Cabal because quite frankly, they are all profiting endlessly from the wickedness, sickness and social divisions in the world as well as from all of the polluting enterprises of global corporations such as the oil industry, the military establishment, of GMO's and Big Pharma. Left unsaid of course is the question as to whether the *real* problem – paired down to its most basic elements – is evil activity being perpetrated on the bulk of humanity by sociopathic elites? Well, we can deal with that problem too in a humane way, but first of all we have to clearly understand the nature and structure of human evil and its embodiment in powerful individuals and institutions before we can begin to consider solutions.

The New World Order (NWO) is described, somewhat disappointingly by Wikipedia as 'a conspiracy theory' in what may yet prove to be a classic – and truly ironic – example of the power of the Cabal to be able to turn what was once a people's initiative into a mouthpiece for the establishment. Because in a 1990 speech to Congress in the aftermath of the Cold War, President George W Bush said:

> *"Until now, the world we've known has been a world divided—a world of barbed wire and concrete block, conflict, and the cold war. Now, we can see a new world coming into view. A world in which there is the genuine prospect of new world order. In the words of Winston Churchill, a "world order" in which "the principles of justice and fair play ... protect the weak against the strong ..." A world where the United Nations, freed from cold war stalemate, is poised to fulfil the historic vision of its founders. A*

250

world in which freedom and respect for human rights find a home among all nations."

Then during the Persian Gulf War, he reiterated:

"What is at stake is more than one small country; it is a big idea; a new world order, with new ways of working with other nations.. the peaceful settlement of disputes, solidarity against aggression, reduced and controlled arsenals and just treatment of all peoples."

Sounds wonderful, doesn't it, but then again, if its only 'a theory' then it appears that George W Bush is in on the *conspiracy* – to pretend it's only a theory – right? Not just President Bush, but Winston Churchill the Chinese Government, the United Nations, and scores of other public bodies and establishment figures down through the years who have specifically quoted 'The New World Order' as a concept where we will have a one-world government – which, as we said in Part One would be a wonderful idea if God for example was in charge. But what if He isn't? Because we're not sure that 1.4 billion Chinese communists would agree to that, nor would a Cabal of career liars and moral deviants either. No, that would simply be too awkward, wouldn't it? So, if it isn't God, or the saints, or even some genuinely decent humans who are angling for a one-world government, then who is? Could it be the sociopaths and psychopaths, the soul-less diabolicals who absolutely WILL keep on pushing for what they want despite the cost, the effect, or the damage to everybody else?

James Warburg was a prominent banker and financial adviser to Franklin D. Roosevelt. James' father Paul Warburg had advocated the setting up of the US Central Bank (the Federal Reserve) in the early 1900's. Given the banking cabal's grasp of all things political and the understanding that as long as they control the money they also control governments, it is notable that 70 years ago James made the following prophetic statement in an address to the US Congress:

"We shall have world government, whether or not we like it. The question is only whether world government will be achieved by consent or by conquest." [cvi]

Well obviously if it's going to be by conquest then we can all wave goodbye to any romantic notions of the Great American Democracy. Let's not forget that 'conquest' does not necessarily mean by force-of-arms, but that it can be achieved by mass deceptions that incorporate the unwitting 'consent' of a gullible, credulous public. In fact, that's exactly what's been going on for decades now in our so-called modern democracies where the façade of democratic elections masks who the real conquest-driven string-pullers and vote-allocators are. The Wikipedia article is actually quite comprehensive in its coverage of the New World Order topic, so please feel free to follow the links and assess for yourself whether we might perhaps be asking the wrong questions here? Because one central premise appears to link all of the so-called 'conspiracy theorists' out there, who invariably point to this-or-that group – or secret organisation or government – as being responsible for all of the world's ills, and, even though they may be correct to have suspicions about the Bilderberg Group; the Rothschilds; the CIA; the Freemasons; the Illuminati; the Tavistock Institute; the Vatican; the Council on Foreign Relations; the United Nations; the World Economic Forum; the WHO; the EU political establishment or the Bill and Melinda Gates Foundation for instance, we must remind ourselves that it is the very essence and structure of many of these often-secretive and opaque organisations that is the real problem. Because as explained in Part One, once any group or organisation drifts away from its humane foundations in pursuit of power or profit, then you can be assured that the tentacles of corruption with all of its wicked perversions will embed itself like ivy around a tree, slowly choking and strangling that organisation until it's the pursuit of corruption itself and all of the lies that fuel, drive and conceal that pursuit that becomes the dominant premise and the organisation's *raison d'être*. Remember? People have consciences – organisations do not. So, instead of looking to wicked organisations that we believe must first be exposed, eliminated and then be replaced... but replaced by what exactly? More organisations perhaps? No. It is absolutely essential that we look at the Covid-19 phenomenon from a perspective of good-vs-evil on an individual basis as well as from a global perspective. That is, that each of us who is genuinely seeking the truth, and who respects justice and humane values, MUST take ownership of the issues and questions before us in our own personal capacity before we can even begin to imagine a collective solution. We also need to be looking at the individual traits and characteristics of those who are driving the agenda and from there, decide whether or not we too want to be conscious partners with evil? Why for example would you choose to sit next to a pickpocket on a train? Why would you return to a restaurant for second helpings of food that made you sick? Why would you bring your car back to a dodgy mechanic who overcharged you the last time? Only a fool does such things – right? Well, why then do we keep on dealing with liars and moral deviants on a daily basis when our consciences and our intuitions keep telling us it's a very bad idea?

It's an incredibly powerful position to hold when you claim the authority to tell the pickpocket that you won't be sitting next to him because he's a thief. Neither will you pay for second-helpings of rotten food. Nor will you trust a crooked mechanic. And, when some authority figure then tries to *order* you to do so; all you have to do is ask them if they are ordering you to be complicit in those unlawful, amoral or criminal activities? What can they possibly say? This is one of the main tactics we use to prevent rogue agents of the State and corrupt office holders from visiting injustices on us – especially in the Irish Courts where it is no exaggeration to describe our collective experiences over many, many years as being the proverbial 'dance-with-the-devil'. A macabre dance where the diabolicals keep changing the tune while tossing us back-and-forth between schemers, tricksters and wily predators, all the while smirking at our confusion, and then charging us for the privilege of being lied to, bullied and robbed under the guise of law. It really doesn't get much worse than being able to name and shame over 40 judges in an open publication, and then telling them (and the world) that we absolutely and categorically refuse to have any dealings with those individuals in their official capacities until such time as the crimes we have reported are properly dealt with.[11] The one exception is when we take out private prosecutions against them when the statutory authorities – as per usual – try their level best to ignore, bypass, circumvent or stonewall legitimate complainants and all the crimes they are reporting.[12] But the interesting thing is that it works! We have now issued around 150 warnings in the form of 'Notices of Proscription & Exclusion' advising criminal miscreants in the pay of the State that they have in effect, lost their privileges to deal with any honest member of the public who so objects. When they try to 'carry on regardless' this provides us with another opportunity to object, to complain, to approach their superiors and to expose their criminal activities all over again. Yes, it takes some effort and determination, but if we were all to start enforcing our absolute right NOT to have dealings with criminal miscreants no matter what elevated titles they posture under – then what can they possibly do in response without risking further exposure not only of their own misdeeds, but also those of their subordinates and superiors who are covering up for them. For those of you who are fed up of all of the lies and deceptions and petty tyrannies being visited on you daily, please take a look at the *Asseverations* page on the *Integrity Ireland* website,[cvii] as well as *The Peoples Tribunal of Ireland* to see how these relatively simple initiatives could be powerful tools in exposing and dismantling any hypocritical hierarchy whose intentions are other-than humane.[cviii]

But, it's not our job today to try to untangle all of the convoluted opinions and possible shadowy players in international conspiracies – or indeed in international plans to reshape the world according to their agendas, because all of the evidence is sitting right there for anyone who has the time or inclination to read through the

---

[11] "Criminality in the Irish Courts and the Absence of the Rule of Law".
[12] "D.I.Y. Justice in Ireland: Prosecuting by Common Informer."

information on the World Economic Forum website for instance – which is one remarkable example of how it all begins with the Covid-19 phenomenon.[cix] It is really quite breathtaking to see how much time, effort and money has been ploughed into the reshaping of our world in every way imaginable, by a group of self-appointed 'world leaders' whose three chief assets seem to be: (i) Money. (ii) Common purpose; and (iii) a sociopathic mindset. Everything it seems, hinges upon three basic premises:

(a) That the world is in all sorts of difficulties, from global warming and climate change to pollution, pandemics and over-population including the imminent collapse of the world economy; and that something needs to be urgently done about it all.

(b) That the peasantry and various backwards-thinking governments around the world are clearly not capable of addressing these problems in a mature, responsible and efficient manner, so instead we will need a one-world government comprising all of the people with *real* power (i.e. money) to lead the people to a better future – one where, *"They will own nothing, but they will still be happy."* (Yes, that's a direct quote from Dr Klaus Schwab).

And (c) that in order to achieve all of this innovation and reformation of society, that individuals must surrender their autonomy to the Cabal in return for an untroubled existence, because obviously, with plans this grandiose and meticulous being rolled out for the betterment of humanity – well, we can't be having pesky objectors or opinionated questioners getting in the way of the Fourth Industrial Revolution now, can we?

Yes, it's a trade off between shutting down your intuition and your conscience and forcing yourself to believe – against all of the historical proofs otherwise – that ruthless, ambitious, sociopaths and psychopaths plus the power-and-profit-driven institutions they have created and installed, really do have YOUR best interests at heart, and, that they genuinely care about our children as well.

'Governance' is an accepted reality in our world, and it is upon that accepted reality that organisations such as the World Health Organisation and the World Economic Forum for example, presume to make their 'governance' projections and presume to appropriate the role and position of governors over the rest of us. But what if the very concept of governance is fundamentally flawed? Now we're not saying that there is no place for governance structures and logistical arrangements for purely practical reasons, but what we *are* saying is watch out for those who would presume to be our governors.

In Part One, we explained how tribes evolved to kingdoms, to governments and to empires; essentially, in pursuit of ever-more wealth and power. We also explained how and why Big Business would want to marry with governments in order to get their greedy mitts on all of that "free" taxpayers' money. That's why it was necessary for Big Business to court or lobby, or bribe or blackmail politicians to get them to implement policies purportedly 'in the public interest' but which really only benefited the Big Boys. All the politicians had to do when approached like this was to continue doing what they have always done so well; just lie to the population, but lie convincingly enough to keep the moolah flowing in the right direction.

But things have changed in the last few years. Instead of going to all of that trouble lobbying, and bribing and blackmailing compromised politicians with all of the convoluted logistics and detailed planning and furtive cover-ups that goes with that, why not just find a couple of generous sponsors who have more money than God, and then the Big Boys can dispense altogether with having to engage with the politicals – unless of course, those politicals are wholly on board with the plan, and are willing to continue deceiving the public for the Big Boys' benefit.

The Big Boys in this respect are the multinational conglomerates especially Big Pharma and the Banks; blended socio-political institutions including the United Nations, the World Health

Organisation and the World Economic Forum; as well as their billionaire sponsors such as Bill Gates, William Buffet and George Soros for example. The International Monetary Fund (IMF) and the unelected officials at the EU could also slip in here as willing co-partners to bring about The Great Reset which is effectively, the application of all sorts of new technology as espoused in the concept of the Fourth Industrial Revolution of Klaus Schwab's invention, which will see people being 'blended' with smart technology in a manner that arguably renders us less-than human. Regardless of whatever exciting possibilities may arise by implanting microchips or enhanced robotics into people, the fact is that once we are blended with 5G or microchips or nanogels, then we are no longer really in full control of ourselves, are we?

You see, the ultimate objective is control, because when you have control, you get your hands on the money – or, vice-versa: if you have enough money nowadays you can effectively buy the control. It's the true definition of a vicious circle complete with all of the vices and the cycles of corruption needed to keep it all turning over until the objective is met. All you need to achieve this overall global control is enough ethical and moral prostitutes in the corridors of power who will willingly sell their souls, or their morals or their consciences for their own personal slice of the poisonous pie – even if it's just a few crumbs from the tyrant's table. Tricks, lies, inducements and punishments; it's all been planned out meticulously, complete with all of the necessarily-misleading sound-bites to convince us otherwise.

"The pandemic represents a rare but narrow window of opportunity to reflect, reimagine, and reset our world." Says Professor Klaus Schwab of the WEC, as if he wants us to actually believe that he didn't even know that The Great Reset and the New World Order was coming.[cx]

This brings us back to the concept of human 'Beasts' as is referred to in the Book of Revelation which, as we explained in Chapter Four could very well be that form of humanity that is operating without that crucial first element of The Triadic Archetype which we commonly refer to as the spirit or soul – our connecting point with the Divine (if you believe in God); and if you do not, then in more philosophical terms, we can call this first crucial element of any truly whole human being, their super-consciousness (or supraconsciousness) a gateway if you like to higher and more noble truths and awareness, sometimes called a moral compass; that silent part of us that resides deep in the psyche and aligns with Universal Law which informs us in a profoundly intimate way of the difference between good and evil.

But, as we explained in Part One and as is shown in the diagrams on p.62 & p.94 even though we must assume that the great majority of mankind is indeed possessed of such an element – a soul or a conscience – that this can and does get affected by all manner of influences including by our genetics and sex (body); by the environment we grew up in, including our education and culture and the often amoral demands of earning a living (mind); and finally, by the personal choices we each make on a daily basis either influenced by a spirit of good, or, in the absence of that 'spiritual' influence, for other less noble reasons.

In this manner, and whatever unfair or disadvantaged cards we may have been dealt in life, we will always have a personal choice to decide between what is right and what is not. It may take some time and effort to educate ourselves as to the intrinsic nature of evil and how to avoid inviting it into our lives, but to coin a phrase, this is no more and no less than the very purpose of human existence; to eradicate evil in its myriad forms from society in order to recreate the proverbial Kingdom of Heaven on Earth, something as we inferred in Part One has not yet manifested itself in human social history. You see, in a world devoid of human evil we would still have free will and freedom of choice, and, because all of us are different and unique, there would always be a smorgasbord of options, and alternatives, and outcomes, and results from all of our endeavours. The only difference is that they would all be varieties of 'good'.

As Ralph Waldo Emerson puts it:

"Sow a thought and you reap an action; sow an act and you reap a habit; sow a habit and you reap a character; sow a character and you reap a destiny."

Speaking of destinies, one question has been troubling me throughout this research. Noticing that some of the key players in the Covid-19 narrative are no spring chickens anymore, I started to ask myself if the thesis that these people are by definition and intent, sociopathic, narcissistic and self-absorbed, then why in their twilight years would they be investing so heavily in world affairs when they might just pop-off this mortal coil at any moment? We know that no-one is in a hurry these days to enter a care home, but with the average age of death being around 83 years, one has to admire the enthusiasm with which these billionaire octogenarians are getting stuck in.

One possible answer is that we have gotten the whole picture completely back-to-front, and the likes of Anthony Fauchi (80), Klaus Schwab (also 80) and Bill Gates (an interesting 66) are in fact saintly philanthropists and visionaries who are dedicating whatever's left of their lives and their fortunes to the public good. Another answer perhaps is that they already have the knowhow and technology to extend their own lives – by selectively 'blending' with nano-technology, by biohacking, or by undergoing organ transplants using cloned substitutes or prosthetic body-parts to prolong life. The almost too-obvious suggestion that they already have an antidote to whatever planned pathogens may be 'served' on the general population is almost too obvious to even mention.

Now, this is only speculation at the moment, but undoubtedly the knowledge is there and, at only a modest $1.8 million a pop for your own personal clone, well, as long as the ethics, morality and legalities are not a problem, maybe we should all be taking out 2nd, 3rd and 4th mortgages in anticipation of the rush to immortality? [cxi] But any such attempts at this sort of 'playing God' by creating

259

amalgamations of human DNA with robotics, cloned prosthetics or nanotechnology – just like some of the contents of the RNA injectibles – risks straying into Frankensteinian territory where our core essence as discrete human beings is altered in such a manner as to prevent any future reconnections with the Original Almighty Divine. In other words, by partaking in any conscious way in the beastly schemes that are being foisted upon us right now on the back of mass deceptions and outright criminality, well, we need to be asking ourselves what this criminal complicity may be doing to our souls?

It is clear to see that evil – even though it has no true role to play in nature – is a fact and feature of human existence to date which, if we allow it to, will not only end up dominating human history from its inception to its almost-inevitable violent and bloody conclusion, but will prevent any future generations from ever realising their individual promise as true children of light. If we allow the diabolicals to fulfil their Machiavellian plans, there is no doubt that we are all headed for Hell – one way or the other – either in this life or in the next.

We have no other moral choice therefore but to resist, and to oppose and indeed refuse, absolutely and uncompromisingly, to participate in the Great Lie. Otherwise we too are complicit.

CHAPTER TWENTY

DEATH BY CONSENT

ADVERSE EFFECTS & THE CHANCES OF SURVIVAL

The data on adverse reactions and reported deaths as a result of the injectibles as opposed to the Covid-19 disease (whatever it really is) has been trickling, pouring and now flooding in during the composition of this book despite all of the attempts by the Cabal to suppress THE final proofs of a truly diabolical agenda which, at very least, is seeking to generate massive profits for Big Pharma on the backs of the likely – and now proven – deaths of hundreds of thousands, if not millions of people. The more sinister projection, that a massive *preplanned* cull of the population is underway may yet prove to be beyond dispute, and we can but hope and pray that we all wake up and see what's happening – and then move as a determined collective to hold the diabolical string-pullers to account – before there's no one left, or indeed even capable, of doing so!

The following April 2021 article has been reproduced in its entirety with the kind permission of independent journalist John Waters – a respected colleague and an erudite, articulate and courageous spokesperson for truth.

*　　　*　　　*

Frontline Truth-telling on Vaccines

Research by Health Freedom Ireland (HFI) suggests a connection between a recent surge in mortality in nursing homes and the vaccination rollout beginning in mid-January. The case is ominous and plausible.

The linked data and graphs [at end of paragraph], prepared by Health Freedom Ireland, would in a functioning democracy be sufficient to bring the current lockdowns to an end, to dissipate all claims of an ongoing pandemic, and prompt a root-and-branch investigation of

NPHET, the HSE, and members of the current government and its predecessor, as well as a total review of the nursing homes industry, which has been the locus of significantly more than half the deaths attributed to Covid-19 over the past 14 months.[cxii]

The Health Freedom Ireland (HFI) research presents data from 60 nursing homes during 2020 and the months of January and February of the present year. This is just a fraction (about one in ten) of the officially recognised nursing homes in Ireland, but the 60 implicated homes appear to account for the preponderance of the additional deaths reported for the months of January and—in particular—February 2021. The 60 homes featured appear to have suffered more than 400 deaths above and beyond the normal even for a typical winter flu season, and this figure would seem to account for roughly 50 per cent of the apparent excess of deaths being claimed for January/February 2021. HFI conducted a study of 10 selected nursing homes in which there had been a dramatic rise in deaths since late January 2021. The graphs and data show an unmistakable correlation between the explosion of mortality and the rollout of vaccination in these homes from mid- to late January 2021.

The HFI summary notes:

'It has been brought to our attention that there has been a sudden and dramatic rise in deaths of nursing home residents since approximately mid-January. Having looked at the data further we believe this needs to be investigated by the relevant authorities urgently.

'On Tuesday 9th March 2021 we notified this situation by email and registered letter to the Taoiseach, Minister for Health, HSE, HIQA and HPRA.'

Failing to receive any adequate responses, HFI then wrote, on March 11th, to all TDs and senators, as follows:

'We wish to make you aware of a very serious rise in the number of nursing home deaths (excess deaths) for the three-week period beginning at the end January 2021. The attached chart (page 3) shows 160 deaths from a selection of just 10 of the 572 HIQA

registered nursing homes that we studied during the period mentioned. In comparison there were only a small number of deaths in those same nursing homes during December 2020 and January 2021. Considering the overall reduction in COVID-19 related deaths, hospitalisations and positive PCR tests for the same period, the dramatic rise in deaths gives us great cause for concern. Please refer to the attached graph on page 4 showing data for these key trends.

'While this sudden unexplained rise in nursing home deaths may be due to a number of factors, we found that the majority of these excess deaths occurred directly after the rollout of the COVID-19 vaccination programme for nursing homes and may indicate a serious adverse drug reaction to the vaccine. If an adverse reaction to the vaccinations has contributed to these deaths, it has the potential to adversely affect the health of all Irish citizens due to the planned ongoing rollout to every citizen.'

As outlined in the HFI letter, the first vaccines were administered in homes on January 11th. Most of the deaths had been attributed to Covid-19, in accordance with standard policy over the past 14 months.[cxiii]

One remarkable graph, showing the levels of mortality in 60 homes since the beginning of 2020, shows fluctuating totals of approximately 100-200 deaths per two-month period up to the end of 2020.

The summer period, July/August, is the lowest at 96 deaths. January/February 2020 shows 130 deaths. The period March/April 2020, which includes a six-week period in which more than 1,000 people were acknowledged as having died in nursing homes nationally, the data for these 60 homes indicates 185 deaths. November/December 2020 indicates 151 deaths. The two-month period January/February 2021, however, indicates that 557 people died in these 60 homes. This figure is more than four times the burden of the equivalent months in 2020; three times the burden in the period of the alleged peak of the 'pandemic' in March/April,2020; and 406 deaths more than occurred in the final two-month period of 2020.

Other graphs show the particulars of individual nursing homes, indicating the dates of vaccination and the number of deaths that occurred post-vaccination. In the various instances, the number of deaths occurring in January/February 2021 is twice, three times, four times, and in one instance 13 times the average rate of death of any two-month period in 2020. Each one shows that the vast preponderance of the deaths in the January/February period—in several cases all deaths—occurred post-vaccination.

This is quite remarkable research, and ought to have been published in every newspaper and on every news bulletin in the country. It has been more or less entirely ignored. Were it to be publicised as it merits, it would all but certainly put a stop to the vaccine rollout.

The HFI research also underlines that, once again, the overwhelming majority of deaths attributed to Covid-19 have occurred in nursing homes. Indeed, the research suggests that the events in the first two months of the current year cause the events of the period March-May of 2020 to recede in significance by comparison.

Similar concerns were expressed by the UK Medical Freedom Alliance in a letter dated February 5th 2021 to Nadhim Zahawi, UK Minister for Covid-19 Vaccine Deployment and Matt Hancock, Secretary of State for Health and Social Care. The letter expressed a particular concern about, *'the impact of Covid-19 vaccines on the very elderly and those in care homes. ONS [Office of National Statistics] data shows that weekly care home deaths tripled in the two weeks between 8th and 22nd January 2021, at a time when there was a massive increase in the rate of vaccinations of care home residents.'*

All this suggests that we have entered Stage 2 of the 'Covid Project' —a designation employed by the World Bank and other international agencies. The first phase was dominated by media-generated mass panic; the second is to be driven by what are being termed 'vaccines', but in several cases are gene-altering therapies, referred to in the pharmaceutical trade as 'disease protocols', which are designed to, in effect, colonise the human immune system and usurp its functions.

As early as April of last year, the Canadian scientist Professor Denis Rancourt was warning about the ominous connection between lockdown and vaccines:

> 'This [the global lockdown] is being done — we're being expressly told — in order to circumvent our own natural immunity, in other words, for as long as it takes so that the pharmaceutical industry can develop, test and distribute a vaccine for this coronavirus — in mid epidemic. So this is the first time in medical history that there is a proposal to halt the natural flow of an epidemic in order to be able to develop and market and distribute a vaccine.'

He objected to this for many reasons, including that viruses have *'co-evolved with living beings, plants and animal, for over four billion years.'* There has been, he said, an *'intimate and even symbiotic relationship between viruses and human beings, and we've always adapted to the genetic shifts of viruses by keeping our immune systems healthy and able to respond and learn the new molecular signatures that it requires to know in order to repel attacks from novel viruses.'*

Instead, we are now being asked to put our trust in mRNA technology, by which a vaccine is designed to mimic the immune system, to trigger the kind of response that would have occurred naturally in the vast majority of people had they simply behaved normally. Professor Rancourt warned that there is no guarantee that this will be equivalent to the full immune system response, which is more or less guaranteed to endure in respect of a particular virus for the remainder of a person's life. No studies had been published which showed an equivalent capacity of vaccines. Vaccination also had known risks: The flu vaccine annually resulted in millions of dollars worth of compensation for side-effects arising directly from the vaccine.

> 'It's an understatement to say, the science of vaccines, and the long-term effect of circumventing your natural responses to periodically recurring virus development is not known. We don't know that it's a good idea or not. We can't fully appreciate all the

intricacies and complexities of that. So I think it's a reckless experiment.'

At the end of 2020, I published an article titled, "The Year of Merciless Killing" based largely on the research of Professor Rancourt concerning other aspects of the Covid-19 'pandemic'. His research, published in a range of papers he had written alone or with others, focussed on the death spikes of the spring of 2020, in particular a six-week period from the last week in March to the first week in May.[cxiv]

By Professor Rancourt's hypothesis, the brunt of the deaths in that period—in numerous jurisdictions right across the world—had resulted from panic and stress arising from the conditions pertaining in nursing homes at the time.

Rancourt used all-cause mortality to analyse the Covid phenomenon, because excess deaths do not lie. He conducted analyses of mortality data in France, Scandinavia, Canada and the United States, and discovered patterns common to all territories that had not been highlighted by anyone else. His investigations conclusively demonstrated that Covid-19 was in no sense a 'killer disease', and that the attendant 'pandemic' has not imposed a death burden in any respect out of the ordinary. The spikes in mortality that had occurred in April had, for the most part, been due to the deaths of elderly people, mainly in nursing homes, whose demise had been accelerated or brought forward by intense stress arising from the policies of the World Health Organisation (WHO) and individual governments acting on its instructions.

In one paper, *"All-Cause Mortality During COVID-19: No plague and a likely signature of mass homicide by government response"* published in June 2020, Professor Rancourt notes a statement by WHO Director-General Tedros Adhanom on March 11th 2020, in which he predicted that the number of cases, deaths and affected countries were about to climb higher. 'Ready your hospitals,' he said.

This, said Rancourt, was 'either the most remarkable public health forecast ever made . . . or something else might explain the sharp peak in all-cause mortality that immediately followed his declaration.' He postulated that the patterns of the 'Covid peaks' — their unnatural height, sharpness and narrowness, their consistency across multiple jurisdictions — resulted from 'an accelerated mass homicide of immune-vulnerable individuals, and individuals made more immune-vulnerable, by government and institutional actions, rather than being an epidemiological signature of a novel virus, irrespective of the degree to which the virus is novel from the perspective of viral speciation.'

He concluded:

'And it happened in exact coincidence and time everywhere. In every jurisdiction that sees this anomalous, unnatural peak . . . the peak started exactly when the pandemic was declared by the World Health Organisation. And the World Health Organisation at that time recommended states prepare their hospitals for a huge influx of people with critical conditions.

'So, the government response to that World Health Organisation recommendation is what killed people, what accelerated the deaths. You can see that in the data, and you can also understand it in terms of how immune-vulnerable people are affected by these kinds of diseases.'

That was Phase 1. Phase 2, which began here in Ireland in January, 2021, is spearheaded by the vaccine rollout.

Since the vaccines have started to come to market, various other dissident experts, including Dr Geert Vanden Bossche and Professor Sucharit Bhakdi, have warned in graphic terms about the likely effects of in particular the mRNA 'vaccines' which both men claim could result in millions of deaths worldwide over the coming years, as covered in this (linked) article which deals with the recent intervention of Dr Geert Vanden Boosche.[cxv]

The HFI data about care home deaths in January/February 2021 resonates with these interventions and dovetails with many other aspects we have become aware of. Yet, neither the expert testimony of world-renowned specialists, nor the verifiable research of Health

Freedom Ireland has been ventilated by the legacy media, which instead publishes or broadcasts daily supposed expositions of 'conspiracy theories' by way of diverting the public's attention from information that might well prove to be life-saving. Another recurring tactic is to appropriate the deaths that occurred in January and February 2021 to refuel the mass panic generators in the hope of duping many more people to unquestioningly accept these potentially lethal vaccines.

The signs already are that in 2021, there will no longer be the necessity to fake the fear porn: The 'vaccines' will do the heavy lifting, and the media's job will be to ensure that anyone questioning the approved narrative, including groups like HFI, will be demonised, marginalised and silenced. Thus, 2021 will become the Year of the True Pandemic — a pandemic of death by pseudo-vaccine.

Writing in October, 2020 to an American editor who had commissioned me to write an article about Covid, I observed:

'I have now arrived at a fairly settled view of the matter, having become absolutely certain that all this was set in train by extremely dark forces with a deeply nefarious purpose. I have not yet definitively settled upon an analysis of what that purpose might be, but it has a great deal to do with establishing new forms of control over the populations of Western societies — of that I have no doubt. The vaccine agenda is undoubtedly part of this, as is the plan to use vaccines to require globalised 'health passport' systems which will be used to force people to accept measures that may be deeply damaging to their health.'

We should be very afraid, also, of the growing desperation of the plotters who, as their plan reaches its denouement phase, will become more and more frantic as the stakes go higher and the risk of being uncovered becomes, in theory at least, more real. How real that risk becomes is, however, a matter of how many people can manage to see through the deception before it grows too late — for themselves or for the rest of us.

It is clear that only the rank corruption of the mainstream media, which daily pumps out propaganda focussing on 'Covid deniers' and 'conspiracy theorists' in order to deflect attention from its collective criminality, enables this horror show to continue. How long more this will go on is hard to say — the signs are that the politicians, health junta and journaliars will now do almost anything to cover up their criminality for a little while longer. But the Day of Judgement is coming, and it will not be necessary to wait for the next life. One of these days, without anyone noticing, the entries on the charge-sheets against these people may morph, as though osmotically, from Fraud to Genocide.

John Waters, May 2021

*   *   *

Two other references of note which are being suppressed by the MSM and by online search engines include Kevin Zucker's ('Klevon') Undercurrents blog and website. Under the heading, "How Many Have Died From COVID Vaccines?" one article states (with full references attached) and here abridged:

- Each year, more than 165 million Americans get the flu shot. There were 85 reported deaths following influenza vaccination in 2017; 119 deaths in 2018; and 203 deaths in 2019

- Between mid-December 2020 and April 23, 2021, at which point between 95 million and 100 million Americans had received their COVID-19 shots, there were 3,544 reported deaths following COVID vaccination, or about 30 per day

- In just four months, the COVID-19 vaccines have killed more people than all available vaccines combined from mid-1997 until the end of 2013 — a period of 15.5 years

- As of April 23, 2021, VAERS had also received 12,618 reports of serious adverse events. In total, 118,902 adverse event reports had been filed

- In the European Union, the EudraVigilance system had as of April 17, 2021, received 330,218 injury reports after vaccination with one of the four available COVID vaccines, including 7,766 deaths

Update: July 16, 2021. 11,000 Americans Dead, 48,000 Seriously Injured as of July 9. The death toll averaged nearly 100 persons per day for the period from April 23 until July 9, 2021.[cxvi]

[The report continues]:
Each year, more than 165 million Americans get the flu shot, and according to the U.S. vaccine adverse event reporting system (VAERS), there were 85 reported deaths following influenza vaccination in 2017; 119 deaths in 2018; and 203 deaths in 2019.

How Many Have Died From COVID Vaccines?
Between mid-December 2020, when the first COVID-19 shots were rolled out, and April 23, 2021, at which point between 95 million and 100 million Americans had received their COVID-19 shots, there were 3,544 reported deaths following COVID vaccination. As of April 23, 2021, VAERS had also received 12,618 reports of serious adverse events. In total, 118,902 adverse event reports had been filed. Data from an investigation by the U.S. Department of Health and Human Services, found that VAERS catches a mere 1% of vaccine injuries, primarily because it's a passive system and reports are filed voluntarily. What this means is that side effects may actually be 10 times or even 100 times higher than reported. We could, in reality, be looking at anywhere from 126,000 to 1.2 million serious side effects, and anywhere from 35,440 to 354,400 vaccine-related deaths. In just four months, the COVID-19 vaccines have killed more people than all available vaccines combined from mid-1997 until the end of 2013 — a period of 15.5 years.

Stunning Lack of Reaction to Mounting Death Toll
Perhaps most stunning of all is that these thousands of deaths and serious reactions are receiving no attention whatsoever. In 1976, the U.S. government vaccinated an estimated 45 million people against

pandemic swine flu. The 1976 pandemic swine flu mass vaccination campaign was cancelled after 53 people died. Authorities decided the vaccine was too risky to continue the campaign. Now, health authorities are shrugging off more than 3,500 deaths following COVID-19 vaccination as either coincidental or inconsequential.

This is 70 times more deaths than the swine flu vaccine, which was halted. If this isn't insanity on steroids, please tell me what is. Maybe murder? This doesn't even include potentially tens of thousands of miscarriages, which is now becoming rapidly recognized as a possible complication of COVID-19 "vaccines."

EU Reports Hundreds of Thousands of Side Effects
In the European Union, we find more of the same. Its EudraVigilance system, to which suspected drug reactions are reported, had as of April 17, 2021, received 330,218 injury reports after vaccination with one of the four available COVID vaccines (Moderna, Pfizer, AstraZeneca and Johnson & Johnson), including 7,766 deaths. Of these, Pfizer's mRNA injection accounted for the largest number of deaths at 4,293, followed by Moderna with 2,094 deaths, AstraZeneca with 1,360 deaths and Johnson & Johnson with 19 deaths. The most commonly reported injuries were cardiac-related problems and blood/lymphatic disorders.

In related news, the Israeli People Committee (IPC), a civilian body of health experts, has published a report detailing side effects from the Pfizer vaccine, concluding "there has never been a vaccine that has harmed as many people." The Committee received 288 reports of death, 90% of which occurred within 10 days after the vaccination; 64% of them were men. This contradicts data from the Israeli Ministry of Health, which claims only 45 deaths were vaccine related.[cxvii]

*   *   *

The third link is more specific to Ireland, and, very interestingly makes reference to the concept of the Covid-19 'vaccines' being used as a Trojan Horse for nefarious purposes, with the link at "www.indymedia.ie/vaccine deaths" saying somewhat prophetically in May 2021 :

> *"The Covid-19 vaccines (aka the experimental gene therapy) are a Trojan horse for two reasons. One is because they have the potential to cause great harm and long term chronic health and even death and the second because they are a key part of "vaccine passports" that are central to forming the global identification and control system of a worldwide totalitarian technocracy. The media, Big Tech and government have made no effort to hide the fact that they intend vaccine passports to be permanent and expanded to all aspects of life from travel to public events, to entry to work, cafes, restaurants and even pubs although these measures will come later once they get the travel part locked in and accepted."* [cxviii]

On the same webpage, there is a short but very succinct 4-minute video clip of Dr Roger Hodkinson of Canada giving us the unvarnished truth. Dr Hodkinson's qualifications, roles and positions as a world-leading expert in the field of medicine, corporate biotechnology and virology are impressive and beyond question. Everyone alive today should listen to his synopsis which is in effect a tight and succinct summary of the Covid-related contents of this book and of the general conclusions arrived at, and is an uncompromising warning of the mortal danger we place ourselves in by swallowing the official narrative and complying with the diktats of the Cabal.

* * *

Finally, we make reference to the sterling work being done by a range of dedicated volunteers and professionals who are amassing all of the evidence necessary to provide incontrovertible proofs of the crimes against humanity that are being committed against us, largely with our own poorly-informed consent. The *'Scientific Expert*

Report on the COVID-19 Epidemic Response in Ireland' will become another powerful exhibit in the battle for the truth, and is available to download on the 'Covid' page on the *Integrity Ireland* website. This is just one snapshot of the enlightening and troubling contents:

Covid19 has a very high survival rate. The high survival rate can be increased further to almost 100% if

the effective medicines mentioned below are used. This survival rate is comparable to a flu season and

there has never been national lockdowns for flu seasons or mandatory vaccines or coercion to take

them in workplaces or vaccine passports to enter bars and restaurants and public events.

COVID19 SURVIVAL RATES	
00-14 YEARS	99.9998%
15-44 YEARS	99.9931%
45-64 YEARS	99.9294%
65-85 YEARS	99.6297%
> 85 YEARS	98.2499%
Jan, 2021	

Sources: CDC in USA, January 27 2021.
https://www.cdc.gov/nchs/nvss/vsrr/covid_weekly/index.htm#SexAndAge
And https://childrenshealthdefense.org/covid-vaccine-secrets/resources

Safe and Effective Medicines - Ivermectin, Hydroxychloroquine, AZT, and Zinc, Budenoside, Dexamethasone, Plasma Antibody Treatment, and Vitamin D are 90% - 100% effective against covid19

Need we say more? Our gratitude to John Waters, Kevin Zucker and to all of the sources quoted for making this crucial data public and available.

For more from John Waters about the Covid-19 phenomenon, please visit: johnwaters.substack.com.

THE TREE OF TRUTH

Of all the consciousness we see
Just a tenth is our reality
And we are left alone to find
Nine-tenths of our subconscious minds.

But if we put aside our pride
To learn those secrets deep inside
We'd find a wealth of knowledge there..
But this is where we humans fail.

For, every glance into our minds, puts previous beliefs behind
Until we come to realise; Divinity resides inside
And patiently awaits the call, for us to come to terms with all
That troubles our so-human minds..

And humbly ask for help from him
Who suffered so because of sin
Who knows our need to be aware
And in His love prepared to share
That knowledge of forbidden fruit
Which Adam stole from the Tree of Truth.

CHAPTER TWENTY-ONE

SYMBOLISM & THE SOUL

WHAT'S REALLY AT RISK HERE

The necessary eradication of evil is the reason why symbolism in particular is so very, very important, because it points us to a state of human perfection that we have not yet realised except in our mythical origins in the Garden of Eden where Eve and Adam, representing the original feminine and masculine in the symbolic forms of the Tree of Wisdom and the Tree of Life respectively, "walked and talked with God, naked and unashamed." And then something went wrong! That 'something' was the abandonment of The Triadic Archetype model by rejecting the Divine Spirit, and Eve and Adam, now reduced to being mere beasts (because they were now only 2/3's whole – or 66.6% of their true selves) had to exile themselves into what then became the Kingdom of Hell on Earth. In this manner, and when associated with key events or themes in history, the number six has become symbolic of 'imperfection' or of 'non-completion' with the number seven representing 'completion' as exampled in the seven days of creation, the seven days to load the Ark "in sevens" (not two-by-two as is so commonly believed), and the seven years it took to build the highly-symbolic Solomon's Temple for example. There are numerous other such examples in scriptures, religions and mythologies and, not-coincidentally, the deep-red end of the light spectrum which is the colour most associated with male heroes and messiahs that measures an interesting 700 nanometres. In this sense, the 700nm compliments the idea of any project being 'completed' once the white (Divine) Spirit inspires the blue (Wise-Feminine) mind to direct the red (Physical-Masculine) body to complete a task. Deep blue, also not by coincidence, registers as 400 nanometres. The same number 40 that represents the feminine theme of rebirth in the 40 days of the flood; Moses 40 days on Mount Sinai; the 40 weeks of pregnancy; or the 40 hours Christ spent in the tomb for example – not to mention literally

275

scores of parallel examples in world history as recounted in 'The Colour of Truth' book. Maybe this accounts for the as-yet unexplained repetitions of these combined colours and numbers not only in key scriptures, mythologies and legends, but also in the heraldry and insignia of nations and empires who tended to align with Truth at any given time in history? And yes, this is all very pertinent to what is happening with this Covid-19 phenomenon, because as long as we remain confused about the nature and essence of evil, we will never be able to get our heads around what is happening and will, either as individuals or as communities and societies, eventually fall prey to its insidious effects.

Now if the non-religious readers out there will please bear with me here as we return to the diagrams on p.62 & p.94 and the notion that human history thus far has been a record of three main themes:

(i) It is the sad and depressing story of the out-of control masculine dominating everything through the misuse of his strength and power, because he is absent the counterbalancing maternal feminine guidance which, in the original model of perfection, would be inspired by our connection with Pure Spirit.

(ii) It is the largely-misunderstood story that is laid out in our mythologies and scriptures of the universe's automatic attempts to correct the imbalances in human society that are caused by the presence of evil. By 'automatic attempts' I mean that the energies of the universe align harmoniously with The Triadic Archetype, and that 'evil' *per se* is NOT a part of The Triadic Archetype model but is in fact the perversion of – and the interference in – the fruitful, harmonious existence of The Triadic Archetype in human society.[13]

(iii) And thirdly, and perhaps most importantly, human history so far is also the story of archetypal numbers, colours, themes and symbols that point to Universal Truth itself, as well as to the

---

[13] The topic of evil and of the Satanic Archetype is discussed in Chapter 13 of "The Colour of Truth Vol 1: Patterns in Light"

presence of human evil wherever and whenever that Truth is absent. But this is a story that has been variously sidelined, misunderstood, suppressed or even dismissed as 'nonsense and fairy-tales' simply because we have all been preoccupied with a physical reality that does not, as a rule, provide us with opportunities for genuine enlightenment. If we think about this for a moment, we can now understand why all of the true spiritual seekers in history – the mystics, saints, sages, prophets and messiahs – all had to go out into the wilderness in one form or another before they could control the body, and calm the mind sufficiently for the Spirit to reveal itself, and so make contact so-to-speak with the Divine. But what is the purpose of any such personal contact if it does not result in tackling THE problem in a substantial way; i.e. the problem of human evil preventing the restoration of the original plan as laid out in the Garden of Eden?

The Cain-and-Abel story has much more to teach us than we might realise, and believe it or not it also has a direct bearing on the Covid-19 phenomenon. Because after the Fall, we are told that Adam and Eve had two sons. Cain was the older son, a big strong farmer, while his younger brother Abel had the arguably more 'girly' job of looking after the sheep while playing his flute. Then Cain and Abel started what we call 'religion' today by making the first ever symbolic offerings to God. This was the first documented instance of a religious ritual. The reason they were making symbolic offerings was because they had been banished from Eden and couldn't therefore connect *directly* with the Divine. So they did the next best thing, they made ritualistic, *symbolic* offerings as an intermediate form of communicating with God – much like you or I would use the mail or a telephone to communicate if we weren't together in the same room. In other words, a transitional substitute for direct, intimate interaction.

The next important part of the story is that for some reason, God accepts Abel's offering but rejects Cain's. But why would God do that? It doesn't sound all that fair, does it? And why isn't there some logical explanation for so many of these otherwise bizarre and befuddling Bible stories which eminent scholars have been hypothesising over for centuries? Well, in case you haven't already worked it out, the answer lies in The Triadic Archetype. Because if we recall, the original model from the top-down is; God, Eve and Adam. Or, Spirit, Mind and Body. Or, Eden, Tree of Knowledge, Tree of Life. Or, Holy, Happy, Healthy. Or, Concept, Plan and Action... and so on. That is the original order of progression. Remembering that this triadic model permeates all of creation, from the properties of light itself to the composition of atoms and molecules, and that 'evil' as a concept is, 'that which does not align with universal laws'; then we can see now why God could not accept Cain's offering, because Cain was in the masculine (3rd) position and Abel was in the (2nd) feminine position, which meant that Cain had to align with his younger brother in the subject position and make his offerings *through* Abel in alignment with Universal Law. In other words, that Cain should accept the authority or supremacy of his younger brother as an intermediary to God so as to reverse the problem that was caused by the Fall when Eve and Adam disrupted the flow from Spirit-to-Mind-to-Body thereby creating the first human 'beasts' who were absent a direct connection with the Divine.

There is a whole other book to be written on the subject of the mythical Fall in context of human history, but it is probably sufficient for now to simply ask the reader if they have any better answer as to why the bulk of the Old Testament is devoted to stories of older brothers being directed to serve their younger brothers and, whenever they do so, good things happen. But when they don't the providence gets set back again; Cain-and-Abel; Noah's sons Shem-and-Ham; Abraham-and-Nahor; Aaron-and-Moses; Joseph and his 12 older brothers; Esau-and-Jacob; Ishmael-and-Isaac; and more such recent examples that occur in the histories of tribes and peoples, and empires and royal houses down through the ages. Here too we see the respective colours and numbers of the electromagnetic spectrum that align with the 700 and 400 nanometre wavelengths of deep-red and deep-blue respectively being repeated throughout

Holy Scriptures and in latter histories in direct alignment with sibling partnerships and messianic missions – as well as in the respective nations they founded – with such startling consistency as to absolutely beg the question as to how so many biblical scholars, theologians and historians down through the ages just couldn't see the wood for the trees?

Perhaps it was because by then, the institution of the Church had become in many ways, anathema to truth itself. It had become a hypocritical hierarchy in other words. Because just like most human organisations down through the ages, it was utterly male-dominated, exploitative and self-serving, with the propagation of the institution of the Church itself becoming *the* priority over any concept of 'truth' or even of faith in any abstract theology. Whether the various saints, scholars and sinners were aware of it or not 'the faithful' were now being managed and led by a Catholic Episcopy who, of financial necessity, were deeply embroiled in power-politics and kingdom-making to ensure the holy furtherance of the gospel.

And even though they couldn't possibly have stitched the whole theory together without some grasp of modern microbiology or awareness of modern history, there are still more than sufficient markers in scripture for any genuinely-independent seekers to have questioned why these providential numbers of 40 and 70 kept repeating themselves – not to mention God's apparent preoccupation with the specific colours of white, blue and red? One must assume though, that even if scribes and scholars had seen those providential patterns they would no doubt have been suppressed by the Church authorities, because having a universal colours-and-numbers theme that challenged official dogma that could be grasped by all nations and creeds would most surely have undermined the stranglehold that the Church has traditionally held on credulous minds. In a way, this also explains why, with the exception of a small handful of individual priests, pastors and bishops who have no doubt been sanctioned and reprimanded for speaking out against the ongoing attempts by the corporate-political establishment to effectively ban public religious practice (now why would they want to do that, eh?); that NO senior Church officials are kicking up a fuss about what's going on. Indeed, the opposite seems to be the case, with the Congregation for the Doctrine of the Faith (CDF) of the Catholic Church – previously known as the Inquisition (yes, the very same) producing this typically-convoluted argument to excuse/permit /allow & indeed 'forgive' (as if they actually have the power to do so) the taking of vaccines that have been developed using aborted human foetuses, stating:

> *"All vaccinations recognized as clinically safe and effective can be used in good conscience with the certain knowledge that the use of such vaccines does not constitute formal cooperation with the abortion from which the cells used in production of the vaccines derive"* … *"the morality of vaccination depends not only on the duty to protect one's own health, but also on the duty to pursue the common good."* [cxix]

I think we should be speaking in plainer terms. The plain fact is that scientists used the bodies of murdered babies to develop these so-called vaccines – and many other injectibles besides – and no amount of cleverly-constructed legal arguments wrapped up in faux-morality or religious doublespeak can justify such barbarism. Just as anyone who benefits from the proceeds of a crime is considered complicit in the original act, we each need to examine our own consciences to see where we stand. Especially in view of the fact that there is NO tangible danger 'to the common good' here other than from the actions of the diabolicals. The Church is siding with the Cabal simply because it is a compromised institution. In fact, there is a very strong argument that inasmuch as any religious institution is more interested in controlling the human spirit than liberating it, then that institution is definitively, part of the Cabal. Just as any organisation will inevitably become corrupted as soon as the existence of that organisation becomes more important than whatever noble ideals it may have been founded upon, one only needs to look at the horrendous child-sex-abuse scandals that have been going on for decades – if not centuries – to grasp the undeniable fact that 'the Church' in its myriad forms and factions has long since abandoned any sense of truth and morality, and is undoubtedly now wedded to – and embedded with – the scheming miscreants who are running the show. In fact, having had personal experience in religious ministry some years back, and with genuine and sincere respect to all decent people who are invested in religion for all of the right reasons, it is perhaps the institution of the Catholic Church, more so than any other organisation in history, that has best perfected the mechanisms and techniques of using mass neurosis and psychosis in the form of so-called 'religious practice' to psychologically disempower and physically control great swathes of humanity down through the ages.

Briefly, on the concept of 'The Messiah' we should understand that messianic missions also align with the Triadic Archetype inasmuch as we have; (i) the Divine Message (concept); then the Mission (plan) and then, (iii) the Messiah (action) who will do his/her very best to teach the message and implement the restorative mission. That mission is one of infusing people with their own individual responsibilities to, (i) rediscover the Truth (the original concept); (ii) to then align with Universal Law (to consciously understand the plan); and (iii) to stop doing or facilitating evil (through actively good actions). Our collective tendency to deify and idolise these historical religious figures is actually an error of understanding, because these central figures are 'only' the vehicles (at best) that try to convey the Divine Message (of Universal Law); and the fact is (as they so often teach us) that any one of us could take up the messianic mission and run with it – or indeed run with any part or aspect of it according to our place and time in history – provided we have the understanding, the wisdom and the courage to act. By all means, we should show these key figures the respect and appreciation they deserve, but this should not get in the way of us understanding the core message itself, otherwise we are just like a class of besotted, admiring students who are so busy conveying our reverence and respect that we're not actually listening to the teacher – nor getting the lesson.

At the risk of being eternally excommunicated or getting a fatwa issued, it needs to be reiterated that religion *per se* is only (at best) a medium through which we attempt to reconnect with The Divine, and that religion would not and could not possibly exist in the presence of that Divine. This is why the religionists amongst us should understand that religion is only a vehicle – a means to an end. To institutionalise it and embedded it in human society in any permanent, dogmatic and hierarchical form is a fundamental contradiction in terms. Much like having massive pharmaceutical companies needing to keep the population sick and hooked on *their* drugs in order to qualify and justify the continued existence of their particular businesses? Surely it is better to have a healthy, happy and holy population (body, mind and spirit) rather than a society that is physically sick and psychologically disturbed, with no sense of true morality; a society almost completely dependent upon 'the authorities' who cunningly shepherd us through life on the absurd premise that they know what's best for us while pretending, like the proverbial wolves in sheep's clothing, that there's no hidden agenda? Again, it's a case of us individuals repeatedly failing to see and understand that we are unique sovereign beings with the spark of truth and life embedded in our far-too-sleepy souls, and that WE – and NOT them – have the inherent authority and responsibility to make our own decisions in life. So why oh why do we keep returning to these so-called 'authorities' who have proven themselves time and time again to be utterly devious and untrustworthy? [14]

Risking the inevitable backlash for bringing up the Nazi's as an example, it nevertheless needs to be asked whether or not all of those otherwise decent, cultured, church-going, law-abiding Germans would have continued to pay their taxes if they knew what Hitler was doing with their money? Or, would they have been so keen to fund a Church that was not only in cahoots with Hitler, but was using their donations to cover up systemic child abuse and to

---

[14] See: "Psychology, Symbolism & the Sacred: Confronting Religious Dysfunction in a Changing World."

protect rapists and paedophiles in religious orders? These are some of the classic historical examples that we use to decry the shocking absence of collective morality amongst an otherwise modern, well-educated population (at that time) and there isn't one amongst us with any sense of decency who hasn't asked ourselves how on earth so many otherwise-ordinary people during WW II could have been drawn into complicity in such shocking wickedness. Well, now we know, don't we? Just look at the Covid-19 phenomenon and tell me, other than the absence of jackboots and swastikas, what's the difference between what the Nazis were up to then, and what's going on now?

Coming back to the symbolism again; it should be clearly understood that just as the archetypical patterns of 'concept, plan and action' or 'spirit, mind and body' are represented in symbolic form in The Triadic Archetype before they can take substantial form, so must any attempts to do global evil be first conveyed via the appropriate symbolism. In fact, just as The Triadic Archetype embodies the 'concept' of truth and perfection at the primary stage, with completion or perfection represented by the number 7 or indeed by three sevens (as the 777 triad of perfection) the opposing 'concept' of evil and wickedness is similarly symbolised in the '666' number of the Beast of Revelation. This requirement; that the symbolism precedes the deed is an historical constant that is reflected believe it or not, in the colours of the flags of the various warring nations down through the ages depending upon which side they were fighting for. The Nazis and their collaborators in WW II for example, opted for the reds and red-blacks of the corrupted masculine, while their opponents almost without exception each sported the combined archetypal colours of white, blue and red (spirit, mind and body).

Similarly with the first year of legalised abortions in Ireland, 2019 saw *exactly* 6,666 babies terminated.[cxx] Just coincidence? Or is this perhaps the symbolic precursor of the horrors to come in a corrupted Irish State which is the pawn of the International Cabal, where interestingly, the phone numbers for all Garda Stations begin with the 666 prefix, and where the HSE's Covid-19 Childline Mental Health Support Services advertise that children should hurry up and apply for the 'safe, tested and effective vaccines' and, should they have any mental-health-related issues (due to the Covid-19 phenomenon) that they can be reached on their dedicated phone line 1800 666 666.[cxxi]

Regardless of all of the convoluted theories of what may, or may not account for the '666' reference in the Bible, it has long been accepted in popular culture that something indeterminable, mystical and opaque links the '666' symbolism with 'that which is wicked', and what I am saying to anyone who will listen is that the Universe (if you like) or The Great Spirit, or God, or Allah or Yahweh or whatever name you choose to use for Ultimate Truth is sending us messages, signs and symbols that we absolutely need to take seriously if we are not to be made complicit in what may be the greatest act of orchestrated evil in the history of humankind to date.

Again, I am in no hurry to deliver silly theories to prospective detractors or fact-checkers, but as I've explained earlier, symbolism plays a massive and crucially-important role in human history despite our general ignorance of it, and we therefore ignore it at our peril. For the doubters amongst us, please read "The Power of Myth" by Joseph Campbell, and you will begin to understand how and why symbolism in our cultures, in our histories, in our legends and mythologies, as well as in our various religions and spiritual practices has such an important role to play in any genuine understanding of the true meaning of life. With this in mind, let's now review a few additional numerical occurrences of the '666' phenomenon which, regardless of one's personal convictions or understandings of what the '666' symbolism may, or may not mean—(and remembering our 66.6% explanation of what comprises a human 'Beast', absent a soul or conscience)—let's see if the following '666'-related coincidences that are associated with the Covid-19 phenomenon may have anything new to tells us? Because it defies rational explanation that these numbers would persist in such specificity other than as the necessary symbolic precursors for the unbridled wickedness to come.

How for example, do we explain the registration of the 2020 'Track and Trace Act' in the USA by Senator Bobby Rush – immediately allocated $100 billion dollars by the US Government – which proposes to give the authorities the power to invade homes so as to enforce Covid tests and vaccinations under registration number 'HR-6666'?[cxxii] How indeed do we account for the almost impossible coincidence that Bill Gates' patent for a system of recording peoples' personal biological data via sensors and such like that are linked to a cryptocurrency system was registered in June 2019 under the patent number 'WO2020-060606' noting that when we put the registration date in numerical form as '06 2019' and add the numbers together we get a total of 18 which also equals three-times-six (666)? [cxxiii]

And how do we account for the alphabetical-numerical values of the names of some of these key individuals – which, when divided by 6 – register these abundant '666' values, other than as some necessary, symbolic demonstration of the evil that is unfolding right before our eyes?

Because we should be in no doubt that the perpetrators of evil fully understand this symbolism and the need for it to be made public so as to 'prepare the way' so-to-speak, for the evils to come. Because one of the rules of Satanism for example which is replicated in a number of secret societies and sinister brotherhoods within the Cabal, is that the participants, or the initiates – or their intended targets or victims – must be made aware of what they are doing. Being 'made aware' of course can be interpreted in many ways. But if the world has already been shown in symbols, colours and numbers that something seriously evil is afoot and we *still* fail or refuse to object? Well, what more can be said other than the quote from Romans 1:20, that (abridged);

"The evidence is all around them, so they are without excuse".

Numerical-Alphabetical Test   *(Please feel free to experiment)*

A	B	C	D	E	F	G	H	I	J	K	L	M
1	2	3	4	5	6	7	8	9	10	11	12	13
N	O	P	Q	R	S	T	U	V	W	X	Y	Z
14	15	16	17	18	19	20	21	22	23	24	25	26

Name	A-N Value	Observation 1	Observation 2
Doctor Anthony Fauchi, NIAID	75+97+48 = 220	Divide 220 by 6 = 36.666...	Divide 36 by 6 = 6x6's = '666' + '666'
Xi Jinping, General Secretary CCP	33+79 = 112	Divide 112 by 6 = 18.666...	Divide 18 by 6 = 3x6 = '666'
Bobby Rush, US Senator (HR 6666)	46+66 = 112	Divide 112 by 6 = 18.666...	Divide 18 by 6 = 3x6 = '666'
'W.H.G.' [Bill Gates] (060606	23+8+7 = 40	Divide 40 by 6 = 6.666...	Initials only. Full name doesn't calculate.
"Build Back Better"	BBB = 2 + 2 + 2 = 6	Small b's resemble 6's	BBB = bbb = 666?
'CORONA'	3+15+18+15 +14+1= 66	Sum total of letters = 66	

Secret Symbolic Triangles: The number 666 is what is known as a 'triangular number' meaning that if we started with 36 units as the base of a theoretical triangle (6 x 6) and then placed 35 units in the row above it, and then 34 and so on until we arrived with just one unit at the apex of the triangle now standing 36 (6 x 6) units high, then the total number of units in the triangle would be six hundred and sixty-six ('666'). Triangles of course are prominent not only in Freemasonry and many other secret societies, but are a staple of any self-respecting Satanic ritual. Whether this has any symbolic meaning in contrast to The Triadic Archetype is a topic open for discussion. But the obvious connections between a dark triad and the personification of Satan or Lucifer in the position of a perverse deity, or 'Prince of this world' who holds dominion over our fallen society are worth a mention in this context, because again, the symbolism of evil is as important as its subsequent manifestation in form and substance. And before people's eyes glaze over at the mention of Satan / Lucifer / Diablos / the Devil we need to briefly remind everyone of two important points: Firstly, that even without personifying it in the form of a scripture-based fallen-angel archetype, that evil is NOT an abstract notion, but is very, very real in our world today. And secondly, that according to Judeo-Christian mythology, that before he fell out with God, that Lucifer was not only the head honcho in paradise, but that he was a very, very clever cookie as well.

Indeed, considering that the roots of The Triadic Archetype are in the properties of light itself, it is interesting to note that Lucifer was the archangel of light before he became an agent of darkness. The transition from one to the other is similar to Adam and Eve's fall from grace inasmuch as all parties sought out, "the knowledge of good and evil" when they had been specifically instructed by God NOT to do so. The separation from the Spirit of Truth in exchange for the sins of the flesh in other words, and then, try and get as many others to join you as you can – through all of the emerging vices of man – so that at least you won't be lonely in Hell.

Noting that the colour symbolism for evil is traditionally red-and-black; red for the masculine energies that are being corrupted by the absence of light (black); it may only be happenstance that practically all of the depictions of the coronavirus are in fact, red-and-black with the spike protein heads in the shape of red triangles? Could it be mere symbolic happenstance too one wonders, that in the process of constructing and engineering the SARS-CoV-2 recombinants (composite constructs that were dispatched by the CDC as samples worldwide) that specific elements of other coronaviruses such as the 'luciferase' aspects known by their scientific terms including, "RecSARS-CoV/SARS-CoV-luc" make inadvertent reference to the angel of light – turned agent of darkness? [cxxiv]

The traditional family unit comprising a father and mother bonded by love is also symbolised in The Triadic Archetype in its comparison to the triangular-circular structural make-up of the atom, with the (red) proton in the role of the masculine; the (blue) neutron representing the feminine and the (white) electron binding them both together in unity and love. (The quarks ✧ define which is which). And for what triangular purpose does the man and woman join in love and affection? To produce children of course, which in turn form communities of interactive individuals and families in harmony with the natural world – just as the individual elements of atoms fuse and combine with other natural properties to create the world we live in.

Thus the masculine and feminine elements of atoms – the protons and neutrons respectively – are bonded together by electrons much as human spouses are bound together in love, in alignment with Universal Law. The colours and numbers too exactly reflect the Triadic Archetype in a field of scientific study interestingly known as 'quantum chromodynamics' indicating the presence of light and colours. The added fact that the process by which atoms form gases, liquids and solids is only by 'sharing' their energy-producing electrons, parallels how love-based family units (represented by white, blue, red) need stable internal unions before being able (under Universal Law theory) to be a properly-functioning part of the greater social community. Because we all know what happens when we try to split an atom – right?

Gender politics and personal opinions aside, this is Universal Law at work, and it needs to be understood as such. The opposite model – one that is absent the bonding Spirit of genuine love has males and females acting as discrete entities – selfish and self-absorbed – 'bumping' into each other in occasional physical-emotional unions (casual sex or obsessive affairs) which, because they are absent true love, cannot possibly remain bonded. As a result, inevitable splits, confusions, distractions, resentments and rejections occur. Very interestingly, this too resonates with all of the sexually-related themes implicit in the story of the Fall, as well as the otherwise-inexplicable heavy sexual symbolism that permeates our scriptures and mythologies. Given the obvious importance of future generations to any long-term plans for society, and given all of the indicators of a preplanned cull and the physical and philosophical destruction of the family unit, perhaps we should be asking why so many world leaders and Covid-19 narrative pushers remain childless? Is this the plan, or the proof, or just mere coincidence? Again, for more on this colour-symbolism topic, the sexual provenance aspect and the manifestation of the diabolical archetype, please look into the proofs and explanations contained in 'The Colour of Truth' book.

A cursory check of '666 and Covid' on Google returned an intriguing list of '666'-named statutory documents, articles, and reported Covid cases from locations as diverse as Quebec, Oregon, Israel, Korea, Colombia, Ireland, Maharashtra, Mumbai, Spain, Pennsylvania, Arizona, Australia… so much so that any discerning person simply MUST ask as to how and why so many apparent coincidences? Other intriguing aspects about the '666' phenomenon in relation to Covid include the required 6 feet and 6 inches (2 metres) social distancing which is a nonsensical measurement in terms of airborne pathogen containment, with the latest restrictions in Ireland for example, limiting outdoor dining to groups of 6 and, provided you have proof of vaccination etc., etc., you may qualify for the newly-introduced 'vaccine bonus' whereby, just as has been predicted, those who are obediently following the diabolical narrative and are complying unquestioningly with the diktats, get rewarded by being let into restaurants and pubs – yes, you've guessed it – in groups of 6. Why 6 again? What's wrong with four, five or seven? Is this imaginary /theoretical/non-existent virus also able to count, and is also aware of all of the demonic symbolism at play, knowing that those who are obediently complying with the 6 ft 6 rule in groups of 6 need not be infected, because they are already in the diabolical bag? [cxxv]

The 'Build Back Better' slogan, first coined by Dr Schwab and his colleagues at the WEC has become the international mantra by which scores of politicians, business moguls and celebrities have declared their alignment with the New World Order. But, just as we keep repeating, the symbolism here is all-important with the three b's mirroring the 666 motif, also somewhat interestingly, in the 'BBB' of the blood-brain-barrier perhaps? Not wanting to stretch the theory too thinly, but in acknowledgement of the possibility that we need to be paying at least *some* attention to these otherwise curious repetitions of troubling symbolism; the fact is that numerous artists and celebrities have been posing quite openly using various demonic symbols such as the all-seeing eye or the statue-postures of Baphomet (Lucifer) including creating an overlapping 666 shape by circling the thumb and forefinger and extending the remain three fingers over the eye, leaving us all asking, if the symbolism isn't really all that important then why are so many of them at it? Happenstance perhaps, or just clowning around? But then again, the theme of influential individuals being secretly or clandestinely united in efforts to bring about social changes that will benefit them personally through various forms of social engineering and political deceptions is sadly, nothing new.

We have already covered enough material to have established that we are facing into a great and foreboding evil. But the more positive news is that we may also be on the verge of a great spiritual awakening. The choice is ours. Will we participate in the Great Lie, or will we chose enlightenment and truth? Choosing the Great Lie is easy. All we have to do is nothing. Enlightenment on the other hand is going to require some personal resolve, and we will have to educate ourselves as to what to be on the lookout for, because most of the evidence is hidden in plain sight, obscured only by the Great Lie, and by our reluctance or unwillingness to challenge and expose it.

Without travelling down unnecessary rabbit holes, let's consider the issue of personal discernment and how we really need to hone those skills so as to avoid falling victim or prey to the guiles of the wicked. In looking at one of the more obvious examples of a movement with all of the in-your-face symbolism, connections and resources to make it a prime candidate for membership of the one-world-government Cabal, the influential UN-based *Lucis Trust* for example presents itself as an altruistic, selfless and even philanthropic entity. But is it – really?

Well, first of all it was founded by two prominent Freemasons in the 1920's and was originally named *"Lucifer Publishing Company"*. Its original New York Office believe it or not, was at '666 United Nations Plaza' and it boasts some 6,000 members, and survives, apparently, completely on donations from a number of well-placed organisations and from big-hearted philanthropists (variously) such as Bill Gates, George Soros, Warren Buffet and the Clinton Foundation for example. It has a subsidiary with the rather cumbersome title of 'The New Group of World Servers' (TNGWS) who are described by the Lucis Trust as, *"Men of goodwill who co-operate.. ..to implement the Plan"* (with a capital 'P').

Two interesting things here: First of all, their line up of 'Men of goodwill' sports as many famous sinners as saints, including Bill Gates, Bill Clinton, George Soros, Aung San Suu Kyi, Nelson Mandela, Pope Francis and the Dali Lama for example (you can decide which are which). Secondly, even if you have never heard of TNGWS, you are apparently 'in' the group if you are possessed of a superior spirituality and are working for 'the Plan'. A Plan first articulated in the TNGWS's *"A World at Risk"* report of September 18th 2019 which advocated, with incredible precognitive accuracy, the need for a global-preparedness exercise (Event 201 happened a month later) and which 'World at Risk Report' very coincidentally, was published less than a week after certain 'highly suspicious' activities at the Wuhan Institute of Virology.[cxxvi] That same Plan apparently is a very lovely-sounding version of a unified world with one religion and a one-world government which they hope to achieve, partly, via spiritual triangulations that ascribe to a sort of universal religion whose inner workings are anything but clear. The Lucis Trust too states that it is:

> "..dedicated to the establishment of a new and better way of life for everyone in the world based on the fulfillment of the divine plan for humanity."

Is the small 'd' in 'divine' here intentional or not, one wonders? All the same, all of this wonderful-sounding rhetoric is very impressive and appealing to someone like me who would genuinely like to see world harmony, truth and justice and all of the noble ideals espoused in their various mission statements. Not to mention the arrival of a great teacher who is a sort of blend of Buddha and Jesus but with the curious name of 'Sanat Kumara'? Mmn, 'Sanat' does bear a passing resemblance to 'Satan' does it not – or is it just me? Otherwise, the narrative is both attractive and convincing. But what if it's not true – with a capital 'T'? What if it's just another aspect of the Great Lie – all fluffed-up and sweet and shiny and wrapped in heavenly lullabies? Because before I would even remotely consider subscribing to any such group, I would do my research and find out why for example Alice Bailey, its spiritual founder, named it after Lucifer in the first place and why they were housed at 666 United Nations Plaza? I would also be asking about her esoteric teachings and about any hidden agendas that the Freemasons may be hiding, such as any origins – commonly-known or not by their membership – in Satanism or diabolical symbolism especially in light of the following quote made by clairvoyant-spiritualist David Spangler who was reportedly the head of the Luciferian Department at the UN (now disbanded) and a stauch adherent of Alice Bailey. [cxxvii]

> "No one will enter the New World Order unless he or she will make a pledge to worship Lucifer. No one will enter the New Age unless he will take a LUCIFERIAN Initiation."

The mix-up between the concepts of God and Lucifer seem to permeate all of the Lucis Trust-related writings and quotes. I would also be asking what it means that the Lucis Trust has 'Consultative

Status with the Economic and Social Council of the United Nations' (ECOSOC) whose US Zip Code just so happens to be '10017'. Intriguingly, the 666th Root of Pi as noted in the Google calculator is: **1.00172029**, and we were informed just today on the news of a great and pioneering mathematical breakthrough in the study of Pi which will be, "of great interest to virologists" it seems. Just mere coincidence again? Or perhaps more cause for concern? [cxxviii]

The relatively recent discovery that 'something highly suspicious' happened at the Wuhan Institute of Virology on September 12th 2019 when all of their viral databases were taken offline is also worth a mention, given that (allowing for marginal time differences) this has occurred *exactly* 18 years to the day (6+6+6) after the 9-11 attacks on the Twin Towers on September 11th 2001 – a day that has already gone down in infamy as the possibly-preplanned precursor to an otherwise indefensible oil-grabbing war in Iraq that killed hundreds of thousands of innocent civilians; the introduction of unjustified terrorism-related social restrictions and legislation that has curtailed fundamental rights all around the world; and the beginnings of a Cabal-driven policy of using shocking events – whether preplanned by them or not – to seize more and more control over the population and over global resources.

That TNGWS "A World at Risk" Report coming out the same week as the 2019 Wuhan panic is probably just coincidence again – even though that Lucis Trust-funded Report, the 1st of a dozen-or-so that are apparently anticipating all of the ramifications and 'necessary actions' affiliated with the Covid crisis – must have been many months in preparation – right? I have no further comment to make on 'conspiracy theories' about 9-11 other than to say that if it was indeed a preplanned, orchestrated act of mass murder by psychopaths in high places, for the singular purposes of seizing power and making profits – well, it would only be par for the course would it not? It would also most certainly align with the Covid-19 phenomenon in the ranks of diabolical actions which are curiously and remarkably linked by '666' symbolism.

Finally, there is the great contradiction between the UN's publicly-proclaimed '17 Sustainable Development Goals' for their '2030 Transforming Our World Agenda' and the manner in which they and their partners at the WHO, including the Lucis Trust's TNGWS and all of the interconnected players in the global Cabal are managing this purported pandemic; which contrived pandemic, arguably, could be the undeclared 18th Development Goal — represented again by 6+6+6? But on second thoughts, maybe there's no contradiction here after all? Because one conceptual way to end war, poverty and world hunger whilst trying to eradicate pollution and improve the environment — not to mention nullifying the effects of any future pandemics or plandemics — would be to radically cull the population and then monitor and control all of those who are left — right?

We simply cannot keep taking things on face value any more when there is so much deception and deviousness at play, and where so much alarming symbolism keeps getting thrust in our faces. The UN-based Lucis Trust may indeed be a genuine organisation with good intentions — despite all of this troubling symbolism — and I would be the first to applaud them if that proves to be the case. But remember what we said about ALL organisations that would attempt to replace our own intrinsic Divinely-inspired authority — not to mention the fact that religion of any sort is only a temporary illusion that points, at best, to the real thing? What we need is less religion and more truth. Less medicine and more heath; and less conditioning and fearful indoctrinations and more true education in the genuine meaning of life. But that begins with me, and with you — and NOT with 'them'. In any event, it is my own personal belief that we needed something like this Covid-19 phenomenon to arrive on the horizon before enough of us would be shaken out of our comfort zones and complacency, sufficient to start asking fundamental questions about what's happening in society; and whether it could be remotely possible that we have all been living a colossal lie? And if so, then what are we bloody-well going to do about it now?

THE TRUTH ABOUT COVID-19
To be understood in context of THE GREAT LIE on p.113

- We are being systematically lied-to by the establishment

- SARS-Cov-2 is not a naturally-occurring virus

- It was neither new nor unexpected

- It did NOT originate in bats and pangolins

- It was the product of unlawful bioweapons research

- It is an artificial, constructed composite of other viral materials designed to efficiently infect humans

- It was, in parts (unlawfully) patented years in advance

- It hasn't been isolated according to Koch's postulates

- It *may* not even be at loose in the population

- It is NOT more dangerous that the common cold or flu

- The samples in existence were engineered by the CDC

- The USA and China have been sponsoring unlawful gain of function research for years

- Face masks are neither necessary nor generally effective

- Airborne viruses in coughs and sneezes can travel 28 feet

- The 2-metre 6'6"social distancing is a medical nonsense

- Social distancing is NOT a disease-related necessity

- It is a symbolism-based psychological ploy to interrupt open, healthy-and-natural social interaction

- Viruses do NOT differentiate between different group sizes

- Biotech companies are not interested in the common good

- Big Pharma is a ruthless, profit-driven industry

- Pfizer had patented the S1 spike protein in 1990

- Beginning in 1989; *"Prospects for even deadlier, airborne microbes have risen because the technology to alter viral and bacterial genes is now fast, easy, cheap, and precise."*
- The PCR Tests are inappropriate for viral detection
- They have been deliberately misused to generate 'cases'
- The rapid antigen tests are not safe or reliable
- The invasive E-O nasal swabs are potentially very dangerous
- Many so-called 'Covid-19 vaccines' are not in fact vaccines
- The injectibles have NOT been properly tested
- The jabs are completely unnecessary as disease preventables
- They contain experimental and pathogenic elements
- They may have been engineered to cause further sickness
- They appear to be attacking fertility locations
- They are known now to have caused many serious injuries, disabilities and deaths
- There has been NO general rise in deaths 'due to Covid'
- Modern vaccines in general, are neither safe nor necessary
- The Covid-19 phenomenon aligns with Satanic symbolism
- The Cabal is using Covid-19 to advance the Great Reset, the 4th Industrial Revolution and the New World Order
- Each of these 'initiatives' require full, fascist-type control of the population by the State or by a One World Government
- Our freedoms, liberties and rights are being erased
- We may never get them back
- There are humane solutions, but we must muster up the courage and the resolve to do what we must without delay or hesitation.

PART THREE

CONCLUSIONS
& SOLUTIONS

CHAPTER TWENTY-TWO

CONCLUSIONS

IS IT REALLY THAT BAD?

In looking back at each of the points contained in The Great Lie and in 'The Truth About Covid-19' lists there are no doubt one or two points that could be challenged in isolation, as the fact-checkers, debunkers and narrative-pushers will no doubt soon be doing, and doing so ardently. But taken as a whole, complete with all of the evidence? No, the Great Lie is a fact of life I'm afraid, and, if truth means anything at all to us, then each and every one of us must work out where we stand on the Covid-19 phenomenon and prepare ourselves for the consequences of taking that particular position. If we go with the flow, there will be consequences for doing so. If we do not – well, we are already seeing what happens to people of conscience and courage who challenge the narrative – are we not? This is not to say that everyone who is railing against the Cabal is doing so for the most principled reasons, but certainly, some of us are, and we may therefore expect the usual punishments for doing so.

For those of us who know how deep, how dark and how despotic the so-called corridors of power are, with all of their corruption and criminal compromises, there is really no option but to question the narrative and to look for the lies. Because you can absolutely be assured that when this many sociopathic minds meet and collaborate, that the lies will be clever and sophisticated and utterly convincing. That is their moniker after all. Lies, deceptions and deceits in the interests of their own perverse ambitions, and anyone who has truth on their radar or who strives to be of good character needs to understand that ANY dealings with the diabolicals in power is an invitation to dance with the Devil. Uncompromising words indeed; but how else do we summarise the greatest assault on fundamental liberties the world has ever seen? The Covid-19 phenomenon may not compare to the bloody horrors and debauchery of Stalin, Hitler and Pol Pot or the ravages of the Great Wars. At least, not yet! But we are still in the very early stages of this carefully-constructed coup, and it may yet be the case that the casualties from the so-called vaccines in particular will rise to levels never previously recorded. Amongst the other casualties of course are the creeping losses of our civil liberties, of the all-too short and imperfect human experiment in democracy, and the inevitable collapse of trust and confidence in our so-called 'leaders' in business, politics, finance, the mainstream media and in the sciences. Unfortunately, by the time that most of us wake up to the dystopian future they have planned for us, it will be too late.

The Covid-19 phenomenon can be described as a great dark ball of shady, complex lies – symbolised by the red-black coronavirus itself, complete with all of its barbs and hooks and sinister beginnings – which only needed that preplanned kick-off in Wuhan to get the diabolical momentum going. Like stupid fish rising to the bait, we have by-and-large jumped wide-eyed and open-mouthed when we were told and snapped thoughtlessly at whatever they threw at us, despite wiser and more experienced others cautioning us against swallowing the narrative whole. Because once they have their proverbial hooks in you – well, it's probably all over. In this case the 'hooks' are physical in the form of the injectibles and the spike proteins; psychological in the fear, panic and bewilderment being caused by the lies; and spiritual inasmuch as we are being told to bow down and worship 'them' instead of relying upon our own, God-given intelligence and intuition. Some lucky ones may yet succeed in wriggling free and may live to fight another day so-to-speak, but most will not. *Their* job (the Cabal and their cronies) is to get YOU to take the bait – either out of innocence and stupidity, or, out of fear and desperation. They are the predators and you are the prey, and choosing to think otherwise can only have one outcome. Those of us who have tried to point out the truth; that whenever predators congregate like this in history, you can be sure that they are eyeing us all up for their next meal; for their sport or entertainment, or simply because ultimately, there's profit in it. And we, the fools and dupes, not wanting to see the dangers out there – and not wanting to accept that we are already like the proverbial fish in the barrel, continue trying to temper our consciences and assuage our whispered intuition that something very dark and sinister is rolling over humanity. All it requires to succeed in suffocating us all, is that we continue to take the bait, as prompted to do so by cunningly-disguised sharks and piranhas out there – the lackeys and toadies and mouthpieces for the Cabal – who have been assured that they too will get their share of the feast once enough dumb fish have been caught. But I would caution those lackeys and toadies; the officials and bureaucrats; the fact-checkers and debunkers; and most of the media mouthpieces, that they too are being deceived as to their roles, and that it's only a matter of time before shark, and piranha and whatever other scavenger species are assisting in the cull, find themselves on the menu as well.

The awful, simple truth of the matter is that those who have unwittingly complied, willing obeyed or just surrendered in dismay and despair to the diktats of the Cabal, have, whether they can accept it or not, contributed to that dark, suffocating momentum and have fuelled the cycle of lies by demonstrating to all who are watching that *this* particular fish at least, is happy enough to swallow the bait whole – and therefore (it is silently implied) you too should feel safe doing so.

Truth is the only thing that will stop them now, and books like this and the messages they contain are just one way that we can try to fight back. Like dropping a large stone into that barrel, it is hoped that the splash and the ripples caused will alert all of those we love and care about – and indeed, anyone who has the ears to hear and the eyes to see, that it's only truth – loud and courageous truth – that can stop them. Truth in facts and reality. Truth in our courage and conviction. And truth in sufficient numbers to give us a sporting chance at least, at keeping our liberties, and indeed our very lives.

And what of those who have already taken the vaccines – in innocence or otherwise? What's to become of them? Well, my greatest regret as we approach the conclusions of this book is that we could not get it done any sooner. I hope it is not too presumptuous to assume that any honest and intelligent person who reads this book will undoubtedly come to the same disturbing conclusions, and, if they have been complying with the Covid diktats and directives thus far and have already gotten vaccinated, well, I hope that the facts and data in this book – as well as the other sources referenced herein – will be sufficient to raise urgent awareness even at this late stage; that further compliance with the Cabal and with their inhumane plans to bring in a fascist one-world government first by stealth and deception, and then by increasingly oppressive methods and means – must now be resisted as best we can, any way we can.

It is not clear at this point whether 'the Plan' will be completed with just one or two injections. It could well be that a series of such inoculations are required over time which, in combination, will acquire the desired results. We can but hope that this is the case, because it gives some hope to those who have already signed up for the vaccinations without knowing or understanding fully what they were doing. The plain fact of the matter is that the Covid injectibles are utterly farcical as a supposed inoculation against ANY purported coronavirus and I remain bewildered and amazed in equal measure at the willingness of so many to partake of something that is 'experimental' (by the establishment's own definition) untested, unproven, and potentially very dangerous!? Don't forget that the bio-medical establishment have supposedly been working on a cure for the common cold since 1965, and they still haven't managed that, have they? Common sense alone should be slapping us around the face for believing that Big Pharma – who incorporate some of the most sinister, ruthless, unethical and amoral conglomerates on the planet – could somehow miraculously come up with an antidote to a new coronavirus—one that was never even properly isolated—and do so in a matter of weeks instead of the usual 7-10 years!

Come on Folks! When stitched together like that it really does look incredible, does it not? So why did we all swallow the bait and trundle meekly to the testing and vaccination centres with sleeves rolled up without a second thought? As we said, there are consequences...

We cannot predict what the death and injury toll from the injectibles is likely to be six months, a year, or even three years from now, but we already know that more people have died in the days and weeks following Covid vaccination in the past year, than have died from the effects of ALL vaccines combined in the last 30 years. That in itself should be a huge red flag. Certainly, more people are getting knocked-off – directly or indirectly – by the combined effects of the lockdown measures (5 times more child suicides); by missed hospital appointments; by cancelled operations; by the unnecessary use of mechanical ventilators; by confinement in flu-infested care homes; and by the after-effects of the injectibles, than were ever going to succumb to a modest strain of the flu, which, even if the illegally lab-constructed SARS-CoV-2 virus is now at loose in the population, is basically all that it is. Some people may say that the figures are all 'relative' and it's all to do with the percentages and the vast numbers being vaccinated. That may be so, partly, but what if you are in one of those percentage groups who sadly 'doesn't make it' – or if it is one of *your* loved ones who is now disabled or in the ER because they too 'trusted the science'?

Because if there's just one reality that we come away with after reading the horror-list in The Great Lie, is that while science itself can be trusted, sociopathic scientists can not. We have to stop making excuses and face the simple truth that there was NEVER at any point in this orchestrated pandemic any *genuine* need to vaccinate anybody. Remember, that the so-called reported 'cases' of Covid-19 were pulled out of the PCR hat, and that essentially *anyone* who presented themselves for a test was likely to be logged as one of the 97% of erroneous results that somehow translated into reported 'cases of Covid' which in turn provided the data for a pandemic. But of course, most of us were not to know this at the time because, again, we trusted what the authorities were telling us and we listened to the mainstream media, as well as deferring to our guardians on social media and their mercenary fact-checkers who were enthusiastically debunking all of our valid suspicions ostensibly, "for the public good" and "in the public interest". But it appears now that we've all been had folks, and for far too many there's no going back. The vaccinated have in fact been turned into walking-and-talking spike protein-creating Petrie dishes in a process that cannot be reversed – and Lord knows what the consequences will be. This is an awful price to pay for apathy and ignorance and for delegating one's authority away. Yet still, on a daily basis the Covid-19 mantra along with all of the reported 'new cases', hospitalisations and Covid-related deaths are broadcast to a largely credulous and frankly brainwashed public who continue to swallow the guff.

It is, as we said earlier, as if we are all suffering a sort of Stockholm syndrome attachment to our abusers and captors. That we would rather continue suffering the lies and the abuse and the serial degradations of body, mind and spirit than to imagine a scenario where we are truly free again. Free to choose for ourselves.

Now I realise that this is a seemingly-harsh indictment of otherwise good and decent people – many of whom I know and care about – who have simply 'trusted the authorities'. It is entirely understand-able why they would want to do so because that's what we've been conditioned to do all our lives. No doubt I would be in the same boat if we hadn't suffered over a dozen years of appalling abuses and betrayals by the Irish establishment sufficient for us to have concluded some years ago, that whatever *their* agenda was, is wasn't good, it wasn't just and it wasn't honest either. That meant that we couldn't be a willing party to it. Coupled with studies and research into so many interconnected fields, and with the shared wisdom of trusted friends, we were in the fortunate position of being able to piece the puzzle together sooner than most and arrive at the shocking conclusion that yes, this is a real global crisis. But it isn't the crisis that we think it is. Yes, it is a cull too, inasmuch as the Cabal couldn't care less how many people will – or indeed must – die in order for the Plan to be fulfilled. And yes, it is also a coup insomuch as all of our democratic establishments and the fundamental human rights that are supposedly the pillars upon which our modern 'Rule of Law' States are founded upon, are – (if all goes according to plan) – going to be absorbed into a global fascist fraternity that will be presided over by a Cabal of unelected, super-rich megalomaniacs and their political lackeys and collaborators, who intend to variously oppress, enslave and disempower those of us who are left.

I do, most sincerely hope that if we are successful in turning this monstrous deception around, that we will all remember the peddlers of lies and propaganda, the faces of the establishment who articulated the lies, and the so-called 'experts' who have demonstrated their expertise so convincingly; that of the wholesale deception of their brothers and sisters; and ensure that they too get their just rewards for having sold out their consciences. Because there is no way that the Cabal could continue the deception without the active and knowing compliance of so many morally-compromised lackeys and toadies in the mainstream media, as well as amongst the ranks of venal politicians and the so-called 'professions' whose only selfish and cowardly concerns is to remain ensconced within the ranks of the Big Boys.

In short, that the Covid-19 phenomenon is the outward expression of an insidious, sophisticated plan, to march us all incrementally, into Hell. But the great flaw in all of this is that they cannot achieve this without our active consent – both as individuals and as a collective. With enough collective consent of course, the remaining independent clear-thinking individuals won't matter. That's why it is crucial – even if we have been consenting and complying thus far, that we all seize this opportunity to say NO!

Not anymore! And NOT in my name!

CHAPTER TWENTY-THREE

COVID QUESTIONS & ANSWERS

So, to summarise. Is SARS-CoV-2 real?

Yes, at least theoretically, inasmuch as it has been unlawfully constructed in laboratories as a composite of various elements of SARS-CoV-1 and other viral elements (remembering that SARS-CoV-1 was unlawfully patented as a composite in 2003).

Did SARS-CoV-2 occur naturally?

No. There is no empirical evidence of any discrete natural zoonotic origins, but there is plenty of damning evidence that SARS-CoV-2 and the S1 spike protein that binds the virus so efficiently to human cells – which can also be used as a 'messenger' vehicle for other injectible 'stuff' – are artificially-created constructs that were deliberately created and engineered, unlawfully, in laboratories that were being funded with private and public money.

Where are all the laboratories and media outlets getting their 'facts' about SARS-CoV-2 from?

From the physical composite samples sent to them by the CDC *et al* and backed up by the WHO in spite of the fact that they have acknowledged (albeit inadvertently) that they have NO SARS-CoV-2 isolates (pure original samples from infected persons).

Is SARS-CoV-2 'a dangerous pathogen'?

No. Whether or not SARS-CoV-2 exists; (i) only in theory; (ii) as a physical reality created from composites that is being stored in labs; or (iii) as a prefabricated virus that is circulating in the general global population; it is no more dangerous than the seasonal flu. It could however, be used by unscrupulous scientists as a vector to deliver potentially-harmful 'stuff' into hosts including mRNA technology that's designed to disable our natural immune systems.

Is Covid-19 (the disease) real?

The Covid-19 *phenomenon* is real inasmuch as the world has been placed in draconian lockdowns because of Covid-19-related *claims* by people who have a vested interest in lying to us. But whether Covid-19 actually exists as a discrete disease with a different set of diagnostic symptoms to those of the common cold and/or flu is largely irrelevant in context of the Covid-19 phenomenon because again, whatever disease is driving this purported pandemic, it is, at very worst, no more dangerous than the seasonal flu.

What are the known numbers and percentages about Covid-19?

In October 2020 at the height of the reported "2nd wave" of the pandemic, with a reported 41.3 million 'cases' and 1.13 million alleged Covid-19 deaths, and, bearing in mind the inconsistencies and general unreliability of the testing-and-reporting methods and procedures already discussed, the reported 3.1% death rate of this world-changing pandemic didn't even approach the 12% mortality rates of the 2003 SARS (No.1) noting also that, *"the Centres for Disease Control notes that 96% of those dying from COVID-19 also had two to three underlying comorbidities. Some may be caused by the virus, such as influenza, pneumonia, and respiratory failure, while others, such as hypertension, diabetes, and cardiac arrest may be unassociated with COVID-19."* [cxxix]

What are the official figures for the severity of Covid-19?

Even with incomplete and selective data, official sources confirm that recovery rates for persons infected with 'the virus' are in the high 90% range, with some qualified sources maintaining that it is 99.98% or better. i.e. similar to any normal cold-and-flu season.

Why did so many countries initiate alarmist and draconian measures up to and including full social lockdowns based upon highly questionable Covid test figures which, in some instances, did not even rise above the standard scientific 'margins of error'?

Official responses to the early test results defy rational

explanation or scientific justification inasmuch as; (i) the testing procedures were inappropriate and unreliable. (ii) The data collection and collation was piecemeal, selective and incomplete. (iii) The various masks-and-lockdown 'responses' were (and still are) wholly out-of-proportion to our responses to other similar infectious diseases and viruses. (iv) Partial or selective lockdowns and social distancing are patently absurd and nonsensical as remedies when dealing with an airborne infection, unless those restrictions are absolute, self-contained and strictly controlled. (v) If we hadn't been informed by the mainstream media that Covid-19 exists, we would have already weathered the worst of it and acquired collective resistance or herd immunity to a phenomenon that would have otherwise been regarded simply as, "a bad flu season" – if that.

Was the virus originally created and/or released in Wuhan, China?

Suspicious activity in September 2019 at the WIV suggests that *something* unusual was going on that *could* account for a lab breach. But even if 'it' can be proven to have originated there, a number of other international players were involved for years beforehand in illegal 'gain of function' coronavirus research that is directly linked to the Wuhan Laboratories. The prior-and-subsequent 'Lock Step' and 'World at Risk' reports, and the Clade-X and Event 201 symposiums; plus the dramatic televised panics in China (rows of trucks fumigating the city) and all of the coordinated preplanned responses since, seem to belie any notion that this was in any way truly 'accidental' or indeed unexpected. In other words, even if it can be proven that 'a new flu-like virus' was somehow released in Wuhan in 2019 and that it genuinely took the Chinese authorities by surprise, that this does not eliminate the possibility that other dark hands were involved in a deliberate preplanned release there that could then be blamed on the Chinese, so as to trigger off the necessary global pandemic and the sequences of world-wide public control measures to follow.

Why has everybody panicked about a relatively mild illness?

Some of us haven't. But what we *are* worried about is the Covid-19 phenomenon and the agenda of the Cabal. Those who did panic were just playing the roles expected of them when trusted authorities decreed that everyone *should* in fact panic, and *should* absolutely worry and fear for their lives. Unfortunately, and very ironically, that trust may end up costing all of us very dearly – if we don't wake up promptly and stop complying!

Have more people died during the first year of the pandemic than would normally have done so?
No.

Why was a pandemic called by the WHO – complete with all of the accompanying social and economic restrictions and widespread alarm – at a point where the number of reported deaths from Covid-19 constituted only 1% of the previous year's flu season?
Good question!

Can ANY respiratory disease be controlled by the establishment?
No. Not once it is in the general population. It must run its course.

Why are the authorities now pushing to have *everyone* vaccinated?
Well, apart from the overall 'control-and-monitor everyone' agenda and the likelihood that the injectibles are being used to deliver other dubious 'stuff' into people for various control and profit-related reasons, there is the huge problem for the Cabal that if a sizeable control group of say 15-20% of the world's population refuses to take the jabs; that as-and-when the evidence of serious adverse effects and deaths of the vaccinated masses continues to rise it will be far more difficult for the authorities to maintain the lie that those vaccine-related deaths are due to some new Covid strains, variants or mutations. Because if the unvaccinated group are not fatally and inexplicably tipping over in similar numbers, well, the mathematics will speak for themselves – right?

Can a virus differentiate between; (i) groups and classes of people;

(ii) between different locations; (iii) between the sizes of social groups; (iv) between those who are seated or standing, or (v) between people who are eating and drinking?

Of course not! What an idiotic, ridiculous notion!?

Does the vaccine prevent me from re-catching Covid?

No, apparently not.

Does the vaccine prevent me from spreading Covid to others?

Nope, apparently not!

Does the vaccine protect me from new variants and strains of whatever-it-is that we are getting vaccinated against?

Nope, apparently not!

So why are we all getting vaccinated?

Very good question.

Can viruses travel more than 2 metres?

You'd better believe it. More like 10 metres in a normal room and apparently all over the place if outside, or if picked up by the air conditioning.

Why has the majority of the public swallowed the official narrative largely without question?

Through a combination of factors which include: traditional trust in 'the authorities'; fear of the authorities; generational social conditioning; believing the MSM; apathy; idiocy; confusion; bewilderment; cognitive dissonance; coercion; various personal compromises; personal and moral cowardice; peer pressure; financial motives and incentives; insanity; ignorance of the facts; and a general unawareness of one's personal right to say 'no'.

But how can the MSM keep on pumping out worrying Covid-19 statistics every day, and have whole news programmes devoted to discussing the *details* of those statistics – complete with experts and politicians – if the whole thing is a massive deception?

Well, if you haven't done so already, then please read the 'CRISIS,

CULL or Coup' book, especially those parts that explain and prove the existence of an interconnected global network of profit-driven vested interests, who are experts in 'social engineering' and who effectively own the mainstream media, the politicians and a great many so-called scientific and medical 'experts' who, like their masters and sponsors, are equally adept at the Great Lie.

So, is SARS-CoV-2 (the virus) the cause of Covid-19 (the disease)?
Possibly, but not necessarily, and only if we accept that both SARS-CoV-2 and Covid-19 exist as 'novel' (i.e. new) entities. Because for instance; if we accept that Covid-19 (the disease) actually exists as a *natural* occurrence arising out of the presence of SARS-CoV-2 in the general population, then either; (i) the various patent holders were trying to *unlawfully* patent something that is natural *before* it was released, or (ii) that they were engaged in *unlawful* chemical and biological weapons research – again, *before* it was released.

Either of these options means that the whole Covid-19 phenomenon is predicated upon serious criminal activity – so why would we engage in that criminal activity by complying with these draconian Covid-19 diktats when doing so in effect renders us complicit in an unprecedented crime against humanity that is making victims of us all?

In what moral circumstances then is it okay to comply?
Self-preservation *could* be a viable excuse for compliance with these diktats if there was a *genuine* threat, but self-preservation (and/or protecting others from possible infection) does NOT even come into the equation when we now have sufficient data to show that there is no unique threat here.

Even in the worst case scenario where avaricious scientists created a highly-infectious bioweapon that has somehow gotten into the general population, it *STILL* isn't mortally dangerous enough to be in the same category as SARS (No 1) or MERS, and is no more dangerous to the general population on average, than a common cold, which we've all lived with for decades without mortal consequences – if not for centuries.

But Covid-19 is a real disease isn't it?

Not necessarily. There is a real possibility that 'Covid-19' (as a disease that is distinct from and different to the usual cold-and-flu season) may not even exist. In other words, that the 'Covid-19' disease may be no more and no less than the usual cold-and-flu season, rebranded under the SARS-Cov-2 name for sinister, profit-generating and global-control purposes, with scientists, academic institutions, politicians and media outlets (who aren't actually in on the deal) believing the contrived 'facts' and flawed 'data' being sent to them by the CDC, the NIH, the NHS, the HSE, the WHO and others; and who are then, unwittingly, spreading the pandemic story with all of its contrivances and exaggerations, as established 'fact'. Even if Covid-19 is something distinct and separate from a standard common cold, it is certainly no more dangerous.

This in turn would account for the official 2020 records that document *less* respiratory-related deaths than in previous years – and no overall lowering of the average age of death within the general population. These are documented, reductive facts that demonstrate that whatever forms the Covid-19 phenomenon has taken so far – other than the health crisis that has been caused by the lockdowns and vaccines – it most certainly does NOT deserve to be called a health crisis nor qualify as a dangerous pandemic.

So is there a health crisis or not?

Yes, a global health crisis is indeed in full flow at the moment. But this is a manufactured crisis of Covid-related cancelled surgeries, of suicides, of depression, of alcoholism, of obesity, of loneliness, of despair, and of thousands of needless casualties and preventable deaths from the very same 'remedies' being imposed upon us by a sinister and diabolical Cabal.

So, are the mRNA Covid vaccines safe?

No, absolutely not. And they're not technically 'vaccines' either.

The emerging data suggests, almost unbelievably, that they are designed to actually undermine (vs reinforce) our natural immune systems, which could have massive and disastrous consequences in the face of new, similarly-structured coronaviruses coming down the tracks. The mRNA injectibles appear to have been designed to actually turn the human body into a pathogen-creating machine.

What's the difference between the injectibles and normal vaccines?

Normal vaccines stimulate the body's immune system to provide antibodies (soldiers with specialised defensive weapons) to attack and destroy invading pathogens such as bacteria or viruses. But pathogens also use their own specialised tools (such as spike proteins) to attack human cells. Instead of using dead or weakened virus samples—(or transferred antibodies from cured people)—to defend against the Covid virus, the mRNA injectibles are delivering spike protein messenger programs right inside our human cells, where (we are told) they are *supposed* to be promoting the creation of antibodies by actually *making* SARS-CoV-2 spike proteins inside those human cells!

Theoretically, this could lead to the possible 'suicidal collapse' of the immune system (as happens with AIDS) in what is known as an ADE (Antibody Enhancement Disease) cytokine storm, which reaction can apparently occur many years after the injection. As well as spike proteins being specially designed to latch onto human cells and thereby cause blood clots and other complications, this may explain the inordinate amount of post-jab adverse reactions and deaths now being reported.

So, is this a preplanned cull of the population?

It's probably too early to say with conviction that the mounting deaths and adverse injuries from the injectibles were 'preplanned' but they *are* happening all the same, and the emerging indicators are that it *was* planned. The Covid-19 pandemic however was most certainly preplanned, and it all appears to hinge on getting people vaccinated above all else. A massive reduction in the population would also exactly align with the long-term eugenicist plans of Bill Gates; with the need to reduce global carbon emissions; and with Klaus Schwab's Great Reset and the Fourth Industrial Revolution.

Is this a coup?

Yes. Absolutely! It's a coup by a network of rich, powerful and influential sociopaths to eradicate modern democracy and replace it with a new model of psycho-biological-political fascism. The most unfortunate part is that most of us are funding and facilitating it whilst being totally oblivious to this fact!

Who profits from the Covid-19 phenomenon?

(i) Big Pharma and anyone who is invested in Covid-related technologies, tests, PPE's, vaccines, signage or establishment literature. (ii) The media organisations being paid to peddle the official narrative *ad nauseam*. (iii) Compromised doctors and hospitals who are receiving payments for towing the line; for reporting the figures and data as they are told; for covering up awkward truths, and for using ventilators and other unnecessary equipment when instructed to do so. (iv) Fact-checking agencies, and (v) all of the multinational corporations who will step in to replace all of the mom-and-pop businesses that have been driven to bankruptcy. And finally of course, (vi) any individuals, agencies, businesses, organisations, financial institutions (such as the banks), national governments and international bodies (such as the UN) who have an eye on global fascism as the ultimate position for them to further enrich and empower themselves.

Who are the biggest losers?

The rest of us. Our children. Future generations. Plus truth, democracy, justice and human rights.

In short, that the official narrative simply doesn't add up in context of a purported health crisis. But it does add up in context of an in-your-face agenda to transfer all power and control to the Cabal, whilst the great mass of humanity, bewilderingly, surrenders its inherent common sense to swallow the lies and propaganda being projected by the UN, by national governments, by the WHO and by the mainstream media as if it is gospel – despite it being the very antithesis of truth.

These few questions alone – along with the professional credentials and personal integrity of many of the responders as referenced in this book – should give us cause to think again, and prompt us to re-examine what it means to be alive at this time in history, where our very futures as communities, as societies and indeed even as a species depends upon the general compliance or non-compliance of the population over the coming months and years. That is, our abject compliance with an official narrative that is increasingly suspect in light of new facts, data, evidence and personal testimonies; and in the face of legitimate, intelligent questions that are being largely ignored, dismissed or suppressed by the medical-political establishment, while their partners in the mainstream media continue to bombard us and bamboozle us daily with false facts and fake news.

So, what's the endgame?

For the Cabal to get every individual on the planet 'marked' and registered on their diabolical database so as to establish totalitarian control over whosoever is left after the great vaccination cull. Then to effectively replace troublesome humans as a workforce with robotics, clones, and genetically-altered semi-human blends of microbiological technology that will inhabit the New World Order as the most efficient, reliable and compliant units the Machine has ever produced: in other words, the full and final establishment of the Kingdom of Hell on Earth – complete with a fully 'marked' and enslaved global population – with 'you know who' and his sociopathic lieutenants in charge.

So, what's to be done?

CHAPTER TWENTY-FOUR

SOLUTIONS

SO, WHAT'S TO BE DONE?

Looking back over all of the information that has been presented, there can be little doubt left in anyone's mind that the Covid-19 phenomenon marks a great crossroads in the story of humanity where, if it already not too late, we need to seize the opportunity to reclaim our inalienable rights and to reimagine society in a format that will NOT reward the narcissists, the bullies, the greed-mongers and sociopaths for advancing diabolical agendas.

The fact is that we already have a One World Government. They just haven't openly told us about it as they develop their insidious networks that incorporate all of the heads of major social institutions, of corporations and governments. We still think we have nation-states, but we do not. Any nation-state that dares to defy the Cabal will soon experience the consequences. Could this perhaps be the reason why five Presidents of poorer countries who variously declared – quite independently, and based upon the evidence at hand – that they would NOT be implementing Covid restrictions or vaccinations, all suddenly lost their lives in suspicious circumstances? We cannot definitively link the two of course other than by random, quickly-suppressed media reports, but unfortunately, when there's so very much at stake then anything it seems, could be possible.[cxxx]

Remembering that many of the central figures driving the Covid-19 phenomenon are highly intelligent sociopaths and psychopaths who can carry out abominable, inhumane acts without any sense of conscience or remorse; and that they in turn are backed by an equally-sociopathic clique of highly-influential, profit-driven sponsors supported by an army of compliant subordinates, employees, bureaucrats and indeed taxpayers who, whether they are fully aware of it or not, are playing the role of Boétie's lackeys and toadies to a truly diabolical establishment. This includes the more obvious players such as the police, the military, the lawyers and the Courts – as well as private security firms and mercenaries – subservient to the Cabal, who can be relied upon to do the dirty and then cover up the subsequent mess whenever the political optics don't look so good. If we begin from that perspective and stop indulging in disabling, romantic notions that, "everything will be alright" or that, "they wouldn't do that – would they?" or that, "there's good in everyone" ..then the rest of us might yet have a chance to turn this thing around. Because, as we said in Part One there's no fence to sit on any more, and those who continue to try to sit on it will inevitably tip into the proverbial 'hell' that the diabolicals have planned for us.

As the Administrator of the *Integrity Ireland Association* I can testify that whilst moaning and complaining about any particular issue may provide *some* temporary relief from the pain and exasperation being experienced, it rarely brings any satisfactory long-term closure. Similarly with our vain attempts to get the authorities to do their jobs and uphold the law and the Constitution; these efforts – which have been multiplied hundreds of times in all of the circumstances one might imagine – have only resulted in serial debilitating disappointments. So much so, that just like any intelligent person would do when faced with a problem that no-one else will solve – one undertakes to do it oneself. And this is the beginning of the journey. To stop seeking out others – even well-intentioned, determined others such as myself perhaps – to be YOUR go-to resource to solve YOUR particular problems. Because the overarching 'problem' *per se* is the very fact that we keep looking to 'the authorities' or to better-positioned others to solve our issues for us.

But the authorities keep failing us and betraying us do they not? So, why do we keep on going back to them for honest answers and humane solutions when it is patently clear that they are neither honest nor humane, and that the great majority of them will flat-out refuse to do the jobs we are paying them to do, if that requires that they challenge the insidious and pernicious activities of the Cabal in any way, shape or form? Career liars, cowards and moral deviants in other words, whose only real loyalties are to their own venal ambitions, and to the Great Lie.

On a more positive note however, hopeful signs are now emerging of a general awakening of otherwise disinterested and disenfranchised people; as to the fundamentals of government and of purported 'necessary governance' and of a global value system that is all about finances, power and greed – instead of true humane principles. Questions that would not even have aroused curiosity two or three years ago are now the topics of strident debates on online platforms – or at least on those platforms where they are still allowed – and there is a growing alertness of all things 'spiritual' as the hypocritical hierarchies that have dominated human history are exposed to the truth; that they are NOT the sources of all wisdom and authority, but are altogether something else – something that is perverse and insidious and diabolical that has no place in any true humane society that aligns with Universal Law.

This Covid-19 phenomenon is YOUR problem as much as it is mine, and if those of us who are becoming aware do not understand this fundamental reality, then the diabolicals will most assuredly capitalise on our ignorance, on our confusion and especially upon our reticence to confront them, so that they can further extend and compound 'the problem' – the very same problem that justifies their existence as purported 'leaders' and 'great thinkers' who have the right and authority to presume to dictate to the rest of us how we should live.

But that authority was not given to them by any Great Spirit of truth and wisdom, but by another party altogether. He who has dominion over all of the Kingdoms on Earth and who can deliver worldly power and riches to those who will simply bow down and worship him. All they have to do is become masters and advocates of the Great Lie. And all it will cost them, is their souls. And what price is that, if they have no concept of morality, nor of truth, nor of conscience!?

For those who have read this far and who still may not be fully convinced of the nefarious culture that infests almost all large corporations, institutions and government departments, we can but beg, plead and cajole that you simply do the research and add up the logic, because every individual who does NOT grasp the seriousness of the situation we're facing is just another stooge – unwitting or not – for the intentions and ambitions of the Cabal.

So, what's to be done then? Well, the good news is that there are solutions to what is going on, which, if we can understand them in context of the Triadic Archetype formula; that is, of the original Universal Laws that we have been systematically violating ever since the Fall from Eden – and, without getting too religious about the whole concept – if we can picture a society whereby each and every human being is aligned with a spirit of goodness; a mind of wisdom; and a body that was devoted to good deeds in the service of all, then we begin to get a picture as to how we might resolve the Covid-19 phenomenon and resultant evils in a positive and constructive way.

But before we get ahead of ourselves, we also need to understand that we are in a colossal battle between the forces of good and of evil right now, and that even though evil has no place in the Triadic Archetype model of perfection it is nevertheless a real and tangible phenomenon that we will all have to deal with before we can recreate any better version of human society. We can't simply 'magic' a beautiful new world into existence in other words without first dealing with the damage that evil has caused as well as the potential for even MORE damage to be caused if we don't get a full grasp of the breadth and depth of the historical problem of human evil, and how it affects all of us who are alive today, and how we each think and act.

If we could use a simple analogy. Imagine that you ('Fred') are approaching my house and you hear an almighty crash-bang-wallop in the kitchen because I have just swept everything off the table – cutlery, crockery, fruit, condiments and utensils – and when you open the door, they are all lying scattered around the floor. I then turn to you and say, "Right Fred. You see this mess on the floor? Well, I'd like you to put everything back on the table please, exactly as it was before I swept everything off!" Naturally, Fred is going to say, "But how can I possibly do that? I have NO idea what the table arrangement looked like beforehand, so unless you have a picture or some such to help me get started, well, I don't even know where to begin!" But then, I tell Fred that it's all about mathematics, and that all he has to do is work out the volume, shape and weight of all of the scattered items in conjunction with the laws of gravity, and then calculate the strength and speed of my arm as I swept the items off the table, and thus, eventually, he should be able to work out where each item was placed on the table before the incident occurred. It might take Fred a lifetime of intense study and of complex calculations before he has any chance of recreating the original layout on the table, but unfortunately that's the task that has to be completed.

The analogy is completed in understanding that the table is the Garden of Eden; the destructive sweep of my hand is the Fall; and the chaos on the floor is the Kingdom of Hell on Earth which humanity has been splashing and thrashing about in for centuries almost completely unaware that this 'reality' that we call human history is at best, a bastardised, dysfunctional version of the original model, which said original, because it only existed in a time-and-place that we have NO personal experience of, we must rely instead on tips, clues and calculations as to what that original may have looked like? But still, given enough information and knowledge, we could possibly work it all out. And that's where universal symbolism comes in. It's the universe's way of directing us where to replace the elements on the table so that we can restart human society according to universal law. The symbolism also serves as a stark warning of what NOT to be doing, which is why we should not just casually dismiss all of the '666' and Lucifer-themed red-black symbolism surrounding this whole Covid-19 phenomenon, but should be both alarmed at its recurrences and alert to the associated dangers and deliver a respectful nod to the Universe as it does what it can to get the human experiment back on track. But because we haven't understood the symbolism all this time, and haven't even accepted that it may be so important; well, we haven't really understood anything of true, essential importance!

The great problem we've had so far is threefold. First of all, hardly anyone has understood that human history is basically, the story of all of us slopping about – largely unaware and unconcerned – in the convoluted mess on the floor, believing all this time that this is actually how it was always supposed to be.

Secondly, that on those rare occasions when an enlightened soul has tried to explain that our current reality (at any given time in history) is very far removed from the 'Kingdom of Heaven' model, they have invariably been ignored, dismissed, vilified, silenced or crucified – in one form or another.

And thirdly, that even though 'the Universe' has been providing us for centuries with all of the information we need to be able to work out the original plan in Eden, that there simply hasn't been enough information or enough knowledge-and-understanding, or indeed, not enough general public interest until the arrival of the Covid-19 phenomenon, which has at last served as a massive prod in the buttocks to those of us who would otherwise have slouched to our graves in blissful ignorance that our lives had been completely wasted in service to all of the wrong dynamics.

In other words, that generations have passed through life each with their individual and collective stories, and that their communities, tribes and nations have done the very same, believing in their respective quests for meaning, for purpose and for essential fulfilment that they were serving the providential purpose, or at the very least, were leading whole and fruitful lives as best they could. But then again, they do say that ignorance is bliss – right?

Not for me thank you! Ignorance at a time like this is tantamount to active complicity in the great wickedness that is being visited upon us. A crime against humanity is happening right now and we simply HAVE to choose a side. Choosing the *right* side will require courage and conviction, and there will likely be consequences for those of us who choose to do so, but really, what's the moral alternative? Because if the prospect of our children being turned into slaves, commodities or guinea pigs for diabolical corporations and governments who will, without hesitation 'cleanse' them from the face of the Earth if that's what's required to turn a profit, well, is that really an alternative?

So, the question now is what practical steps can we take to address the Covid-19 phenomenon and the purported New World Order that will, if it is allowed to manifest itself, be the culmination of the powers of darkness, of deception and of evil over a humanity that was originally designed to embody the Almighty Divine?

*   *   *

Well, first of all let's look at our positions as individuals in alignment with The Triadic Archetype perspective, beginning with the spiritual-moral-conscience-based approach which will not necessarily require any application of religious principles *per se* other than the traditional practices used by sincere truth-seekers, sages and mystics to access the higher realms of consciousness. This place of supra-consciousness is arguably where all of us can connect with what we have called 'The Divine'; that spiritual component of the Triadic Archetype that informs and directs our minds and intellects to do only that which is good. Leaving aside all of the possible arguments about individual concepts of 'right-and-wrong' or of subjective perspectives or objective opinions, the main point to grasp here is that if all of us had sufficient knowledge and awareness there would be very few disputes about what was, and what was not, essentially 'good'. But the journey of awareness begins with a commitment to first seek out the truth and then resolve to make whatever changes are necessary in our own lives so as to accommodate those truths as they are revealed to us, and thereby, at least in our own individual life-spaces we make truth manifest in thought, in word and in deed. So, the very first thing all of us need to do is make a commitment to goodness, to truth and to integrity – meaning, "Being the same on the inside as we appear to be on the outside". This is not to say that we will become saints or sages overnight, but we must really start to make concepts like 'truth' and 'sincerity' and 'integrity' the core essential values that we use to evaluate all of our daily dealings, and, when-and-if we come across people or situations that challenge those core values, well, there are certain things we can do about that too without compromising our core values and without participating in any such wrongdoing. (More on this in a moment).

We also need to make 'concern for others' a priority, no matter the perceived diminished monetary value of those particular persons in a perverse system where everyone is measured and evaluated by their financial potential. My own son Danny – a special needs boy – would have no inherent 'value' in any such New World Order system. Are they telling me that the source of all my daily joy and that of my family, and the reason that we have all developed an acute understanding of the true value of things that are immeasurable in monetary terms; that Danny himself has no value? How preposterous and perverse! There is hardly a soul who has met Danny at soccer matches, in local shops or even in the Irish Courts (where he functions as an untouchable and therefore very efficient bodyguard for his Dad) who cannot say they weren't touched by his innocence, his personality, and his dedication to his father. But in the new dystopian future, anyone such as Danny who is perceived as having 'no value' in a commercial sense can be disposed of. This is already happening with numerous Down's Syndrome babies being aborted by prospective mothers who either have no appreciation of the true value of the challenge of raising a special needs child, or, whose personal values resonate around all of our inherited-and-false perspectives of what true 'value' and true 'worth' means in a diabolical world? Again, could it possibly be mere coincidence that in 2019, during the first year of legalised abortion in Ireland we clinically 'disposed of' precisely 6,666 babies?[cxxxi]

A senior Executive in a UK-based special needs charity raised her serious concerns when she noticed that twice as many special needs people had died since the advent of Covid than would normally have done. Upon investigation, it was discovered that the UK Government had apparently committed an inadvertent 'oversight' in issuing DNR Notices (do not resuscitate) for people with special needs that had been diagnosed with Covid. 'Oversight'? Really? Is that credible? And if so, it is manslaughter or even murder by oversight – is it not?

You see, if we can accept that the problem of human evil in its various forms including ignorance, apathy and abrogation of moral responsibility is at the heart of all of this unhappiness and dysfunction in human history, and if we then understand that the problem is very greatly compounded and multiplied when evil individuals are given licence and authority to govern, direct or lead us, then we begin to understand that any honourable version of a future society that aligns with Universal Law will have to have filters in place that will prevent sociopaths and psychopaths from holding any positions of power and authority that would facilitate the further expansion of evil in society.

So, rule number one is that we each make a conscious decision to align ourselves as best we can with the universal concepts of truth, honesty, knowledge, sincerity and integrity, and then wrap all of these qualities up with compassion for others (yes, even for the diabolicals) as we evaluate everything and everybody around us. Now when we say 'evaluate' yes, we are telling you to start actively making judgments and to apply your knowledge and discernment. The oft-misquoted Bible verse, 'that one should not judge others lest thou get judged oneself' is a mistranslation of, "do not *condemn* others lest thou condemn thyself". Because we all have to make scores of judgments every day do we not – some more important than others – otherwise we would have no control at all over what happens to us, would we? But condemning someone is a different concept altogether because there is no returning from a condemnation. So, the admonition not to condemn others is a direction that we should always leave the proverbial door open for people to repent, apologise or make amends for any wrongs they have committed. This is why, even with all of the apparent justifications in the world for doing so, that those of us who genuinely understand what evil is and how it operates through human agency; that we must never presume to condemn evildoers in a manner where there is no prospect of their personal redemption. No Madame Guillotine in other words. We should be judging and assessing them, absolutely; and we should continue to take whatever protective or defensive actions we need to protect ourselves and innocent others. But we must never condemn the wrongdoers to eternal Hell so-to-speak because that is not our prerogative. They may choose to go there of their own accord, but that is their concern. We on the other hand must never, ever follow them into Hell, and this needs to be understood literally as well as theoretically. Unfortunately, that is precisely what we've been doing every time we listen to the lies or believe the propaganda or fund the schemers-and-deceivers, and we really, really need to let this fact sink in, and sink in hard and deep.

Remember the pickpocket and the dodgy mechanic and the chef that tried to poison us? Well, whose stupid fault is it if we return to these people and allow them to continue to rob us, to poison us and abuse us? "Fool me once, shame on you. Fool me twice, shame on me" – right? The last line that's missing from this maxim is, "Fool me all of my life, and I probably deserve to join you in Hell as well!"

But what are we supposed to do when our political leaders, our big bosses and the establishment in general is comprised almost exclusively of these amoral, diabolical types? How are we supposed to live, or pay our mortgages, or raise our kids properly without making individual moral compromises that effectively keep these people at the top of these long-established hypocritical hierarchies? How do we avoid falling into the, "I'm just doing my job" category of knaves, villains and reluctant compliants in the great historical conspiracy to oppress, defraud and disenfranchise the bulk of society?

Yes, these are the million-dollar questions are they not? Tough questions too! But who ever said that this was going to be easy? For those who are already embedded with the establishment in their roles as civil or public servants for example, the transition from being a Cabal-dependent complicit to being an empowered advocate of truth and right principles may require a complete rethink of one's career path and whether or not it is truly worth one's soul? Because whatever hardships that arise from doing what is right, they pale in comparison to the eternal consequences of choosing the alternative, because choosing good is always, and invariably, worth the effort.

Two examples of my own experience spring to mind. The first was a Garda who had a genuine interest in the youngsters in his local community and was trying to set up apprentice schemes and a youth club to give them a decent start in life. Seen as a bit of a goody-two-shoes when he sought official funding for these projects, he was dismayed when a jealous superior tried to re-post him to an isolated rural area. So, he resigned from the force and continued his charitable youth work instead. Blessings came his way and he became a successful property owner who could then independently fund the various youth-related initiatives which his superior in the Gardaí so desperately wanted to fail. He also set a strong moral example for anyone facing similar decisions in life.

The other example is that of an Irish soldier who, when being ordered to enforce Covid mask-wearing at a local shopping mall, advised his Commanding Officer that he could not, in good conscience, be party to orders that he believed to be amoral and wrong – and indeed criminal in essence. An avowed atheist, this man is an example to many religionists as to what a moral compass looks like. And yes, he too resigned from the Army and is now devoting his energies, along with his partner who is one of those rare creatures – a lawyer with a conscience – into tackling the Covid-19 phenomenon from a moral-and-legal angle.

Very much like the *Integrity Irel*and journey in challenging and confronting systemic wickedness and injustice in the Irish justice system, unexpected blessings have come our way as well during our pro-justice journey, not to mention some truly incredible 'coincidental' protections of myself and my family that just so happen to align with the Triadic Archetype. These protections appear to have come about in response to our belief that if we simply strive to do what is right and we continually keep educating ourselves as to what that actually means in real terms – and as to what is required of us to achieve that outcome – that the laws of the universe will support and empower us as soon as we make the decision to act.

The concept is actually very simple. Just as a tall man must bow his head when walking under doorways if he is to avoid banging his head; so must we all try to identify natural and universal law and align ourselves with it in our daily lives. Because once we are aligned, the laws themselves will protect us from stupidly banging our heads off low-hanging doorways all the time. But if humanity continues down this path of ignoring Universal Law by failing to grasp the essence of human evil and failing or refusing to do anything about it, well, then it's only a matter of time before we all succumb to the damage being caused. Because make no mistake Friends, what we are experiencing today is a repetition of all of the dark and sophisticated scheming of the Nazi regime, complete with state-controlled propaganda and the suppression of individual fundamental rights, as the population of the world is tricked, deceived, bullied and intimidated into being party to the greatest act of physical, psychological and perhaps most importantly spiritual genocide the world has ever seen.

And just in case we haven't put that clearly enough for those who may still doubt the reality of what's happening, yes, we are telling you that if, after reading this book you still think it's okay to believe the official narrative; to support and fund the Cabal; and to comply with all of the diktats and directives without question, then yes, you too are now complicit in the globalist's lies and deceits; in the move towards a fascist totalitarian society; and as an active adherent of the darkest aspects of human nature that will undoubtedly result in the planned extermination of millions of innocent souls.

There is a strange insanity at play when millions of people accept layers of officially-sanctioned lies-upon-lies without questioning their own sovereign ability and personal responsibility to resist the deceptions and discern for themselves what is essentially true or false. We have to lose the blinkers, and we have to take ownership of who we really are in these strange and unusual times. The Covid-19 phenomenon may be a disaster for most of us, but it is also an opportunity to be part of an historical occasion for change, and for social reformation, and for reinvention of the human experience.

Still, the pressing question remains. Where do we all start with this rejuvenation of moral perspectives and the reinvention of decent society? Are there any practical steps we can take to confront evil in its various forms as it is manifesting itself today in the form of the Covid-19 phenomenon? In other words, can I as an individual do anything at all about what's happening, or is it too late already?

No, it's never too late to choose the truth!

CHAPTER TWENTY-FIVE

THE POWER OF SAYING NO!

USING THE TOOLS WE'VE GOT

So, what can we ordinary people do? Well, the answer is as simple as it is profound. Just say "NO!" Say 'NO' to lies. Say 'NO' to inequality and injustice; and say 'NO' to anybody and everybody – no matter who they might think they are – who tries to pull, push, bully, direct, order, tempt or cajole you into Hell. Because the alternative is simply unacceptable is it not? To condemn our children to futures where at very best they will be the 'educated' (trained, conditioned and indoctrinated) slaves to a diabolical Machine with no prospect ever, of becoming True human beings who are healthy, happy and truly 'whole' (holy) in the original 100% meaning. Instead they will be some sort of captive hybrid human that is part biotech components, part-automation and part trackable software with irreversibly-altered DNA, programmed to target fertility, lifespan and disease sensitivity; an other-than-human being that can be completely monitored and controlled – and possibly even cloned – to provide the most efficient product for the Machine and for the psychopaths in charge. In time, those of us who continue to submit to the demands of the Cabal will almost certainly lose any capacity for genuinely independent thought or personal spiritual enlightenment such as is necessary for us as a collective to realign human society with Universal Law.

Mightier Than the Sword!
But there is something that we can all be doing about the Covid-19 phenomenon in an urgent and immediate way. In fact, the encouraging truth is that we are already beating and defeating the diabolicals in various ways just by saying 'NO' – and by saying it emphatically in various different ways that are as effective as they are audacious! Now, this may only be happening in small and isolated pockets at the moment, but it IS happening and it IS real,

and the templates and tactics we are using to achieve this are available for everyone to use anywhere in the world. All you need are some determined minds who are ready to use these tried-and-tested methods in your own location, and the Cabal WILL ultimately fall! Everything being proposed here is absolutely lawful according to the law of the land, but more importantly, the nullification of systemic, institutionalised evil in each of our lives and communities is a requirement and responsibility that accords with Universal Law. So, with 'You-Know-Who' on our side, well, what's stopping us eh?

This first 12-step process, we call the 'HAFTA' procedure; (Holding Authorities Figures To Account) because it "hasta" be done by us, otherwise the corruption and criminality will never be addressed. 'Hafta' also happens to be the slang word for 'graft' or 'bribery' in the Hindi language, as well as meaning 'week'. This 12-step process has been developed over many years in an effort to empower the little guys who come up against the Cabal, and it can be used against anyone who has committed proven offences against the public. The regulations and laws may be different in other countries of course, but usually, we can find similar methods by just educating ourselves about the options. All you have to do is follow the same formula and you WILL put a stop to the lies and deceptions and abuses going on.

The major problem to date, is that very few people have realised or understood that each of us can take these direct actions either as individuals or as a collective, or, maybe those who *do* know have balked at the sheer audacity of telling a judge for example that he is violating his oath of office and WILL be placed under arrest if he continues to do so. This is one reason why we hope to bring out another short guidebook about the HAFTA process under the catchy title of: *"JUDGE-HUNTING – & Other Fun Ways of Protecting Our Fundamental Rights."* But for today, we are just going to summarise the 12-step HAFTA concept with a couple of clarifications before we begin, urging anyone and everyone out there who sees merit and possibility in these ideas to please replicate them in your own vicinity and just see what happens when you do.

First of all; Notices of Proscription & Exclusion: These are being used with considerable effect on persons in the employ of the State who are committing crimes on the people under the guise of their official roles and positions. These Notices are being used to name, shame and blame miscreants in the pay of the State, and, when the authorities – as per usual – do everything they can to avoid holding these people to account, then we simply "Proscribe" them! In more simple terms, we basically tell them that until such time as their crimes have been dealt with properly according to the law, that we – as law-abiding citizens / residents / honest people – cannot have any further dealings with them for fear of becoming complicit in their crimes. We then publicise the Notices as a warning to the public that dealing with these people in any official capacity could very well be unlawful, and that they would be best advised to write to that person's superior (if they too are not on the list) explaining why we cannot lawfully have dealings with persons who are engaged in routine criminality. This is a very powerful way of simply saying NO that has reduced the number of corrupt civil servants, gardaí, lawyers, judges and politicians who can mess with us, and who would otherwise continue their underhanded and insidious activities against innocent and unwitting members of the public – with impunity. For the tactic to really be efficient we will need more and more people to grasp the simple concept; that if anyone lies, cheats or deceives us, then they immediately lose their privileges to deal with honest members of society. Period!

One anecdote springs to mind in context of the 'involvement' (by association) of the then Taoiseach of Ireland in a vicious campaign of harassment and intimidation of our family. On the occasion of our first and unexpected (for him) meeting in a café in 2016, and as he rose with a plastic grin on his face to extend his arm for a handshake, I told him firmly, *"Sit down Mr Kenny. You don't get to shake my hand!"* The look on his face was priceless. Now this may seem somewhat rude or impolite at the time, but it wasn't actually. It was simply me being honest and refusing to play a farcical and hypocritical role as an overawed peasant who was in the presence of a supposed political giant. I was basically telling him NO! You don't get to play your devious games with me Enda. I then gave his sidekick – the then Minister for Justice Frances Fitzgerald – a short but hopefully educational talk on of the concept of moral turpitude, and how her obvious expertise in that arena was nothing to be proud of. The plain-clothes Garda bodyguards then realised that their precious charges were apparently choking on their doughnuts as we took our leave with both Kenny and Fitzgerald trying to hide their embarrassment with furtive glances at the bemused patrons, and nervous half-giggles.

You see, when we show these people that we can see through the lies and pretences, they really don't know what to do with themselves other than run and hide in shame. Not out of any remorse that they have done something wrong, but because they will always be shamed in the uncompromising presence of truth.

Generating 'Asseverations' through the *Integrity Ireland Commission* is another powerful initiative that could be replicated all over the world by following the same processes we have done. Asseverations are legal documents that have their own intrinsic authority and jurisdiction simply because they are entirely composed of written, positive law. The simple idea behind the creation and production of I-I Asseverations is that they remove all of the confusion and obfuscation and outright lies that one is likely to encounter in a typical Irish Courtroom for example, and – from what we are hearing – this particular problem is not just confined to the Emerald Isle. What we've been doing with these Asseverations is raising simple but explicit questions of law and researching existing texts that pertain to that one question (through a simple online search) and then put the results in order of precedence.

For example, if we were to ask the question, "Can the State force me to take a vaccine?" ..then we may unearth several different laws and principles on the subject including international law, EU law, the Irish Constitution, Acts & Statutes and so on, all the way down to dictionary quotes or the recorded opinions of individual judges or lawmakers. We have already identified over a dozen discrete types-and-levels of law arranged in order, on a simple one-page text template that anyone can use for any question that is of interest to them personally. Once they have completed their research (as best they can) then the results are returned to the *I-I Commission* for a 'jury' of researchers to assess, add, or amend the research document until we arrive at a legally-specific answer to the original question that contains NO personal opinions or other unnecessarily-distracting narrative other than whatever editorial text is required to made the document readable and easily understood. The resultant Asseveration which is never more than four pages long and which will always have a simple one-page, question-and-answer version for people to download and print out for themselves, can NOT be challenged by any lawyer or judge for three main reasons: [cxxxii]

- Because the Asseveration document only contains quotes from existing 'positive' (written) law – the very same law that our governments and Courts claim to honour and respect. So, there's nothing they can add, and nothing they can take away from that text. They simply MUST accept the Asseveration text as-is, or be in direct (criminal) contempt of the law itself.

- Because we have established the foundations of the *I-I Commission* over many years by 'serving' specific legal documents and Constitutional Notices on all of the Irish statutory authorities, duly copied to the European Union, which clarified all of the legal constructs and the reasons why the people have *had* to set up these independent initiatives because of the *proven* absence of the Rule of Law in Ireland. [cxxxiii]

- Because if any judge, lawyer or politician does try to challenge an Asseveration, they must then deal with the *I-I Commission's* foundation documents which in part, name over 100 Irish officials and office holders, including some 40 judges, in proven criminal activity, as well as somehow account for why the Irish authorities are engaged in a documented criminal conspiracy to ignore, suppress and cover-up those crimes. [cxxxiv]

Whatever set of legal-related circumstances you may be facing, an Asseveration, properly done, will bring clarity to the situation and will prevent rogue 'Officers of the Court' from deploying all of the usual tricks and deceptions that are designed to part you from your money or to frustrate the prompt and legitimate resolution of your case. These Asseverations should be done *before* you go into Court, or, if it's a Covid-related issue, you simply carry a copy with you after serving the same on the respective authorities, knowing that none of the so-called 'statutory authorities' all the way to the Dáil, the Taoiseach, the President or the Supreme Court – that NONE of them can, or will be able to challenge those Asseverations, and if they dare to do so, well, then we'll have Constitutional grounds to appeal. They will now be facing the *I-I Commission* in a public Court; and THAT is something that was taken off the menu when we received an unsolicited letter from the Four Courts in Dublin informing us that Gardaí and Court Security had been directed NOT to allow us access into the building. So much for public courts eh? See, we ARE winning!

The Peoples Tribunal of Ireland (PTI) is the other important new initiative that is founded upon the same originating documents as the *I-I Commission*, but the PTI is taking a more comprehensive approach to the systemic abuses of power and authority here in Ireland, by lawfully establishing People's Courts that will operate in practical ways whilst exercising Constitutional authority and jurisdiction over any arm or agency of the State that is operating in open violation of the law – something that practically all of them are at. The resulting PTI Determinations will, according to the letter of the law, have jurisdiction over the Supreme Court of Ireland based upon the text of Article 38.3 of the Constitution which allows for the establishment of 'Special Courts' (in this case the PTI) in circumstances, *"where the ordinary courts are inadequate to secure the effective administration of justice."* This fact too has been established beyond question and without any legal challenge from the whole of the Irish Establishment and the EU Parliament, and the PTI was inaugurated on July 1st 2020.

This brings us to the use of 'common informer' prosecutions which date back to the 1850's in the British Empire and should be available in similar form in all common law jurisdictions. That would include previous British dominions such as Canada, Australia, New Zealand, USA, India, Pakistan, Kenya and South Africa for example – as well as in the UK if they haven't already been conveniently abolished. Very simply, a 'common informer' prosecution is when you or I take a complaint directly before a District Judge or Magistrate using a single sheet of information. There's no need for a solicitor or barrister. There's no requirement that you inform the police or the DPP. You don't even have to tell the Courts Service that you are coming. And if you establish that a crime has been committed, then the judge HAS to issue a summons for the accused to attend Court, and YOU then get to prosecute THEM. And the best part probably – as we explain in the *D.I.Y. Justice in Ireland* booklet – is that the process is free! [cxxxv]

One of the reasons that we can confidently claim that the Irish domestic Courts are, "inadequate to secure the effective administration of justice" is because we have lodged several of these applications for criminal summonses against some of the biggest names in the Irish establishment, and all we get are papers going missing, hearings being inexplicably erased from the records, no-one responding to our follow-up calls and emails, and judges fleeing their Courts in such haste as to prompt us to enquire as to whether there may be something wrong with the food perhaps m'Lud? It remains an open indictment of a criminally-compromised justice system that even after Supreme Court Rulings have endorsed the validity of the common informer process, that Irish Supreme Court judges continue to conspire with their colleagues to try to ignore all of those valid applications – some of which have now been gathering dust for months and years. But before anyone thinks of saying it – NO – this is not at all a wasted exercise by us, because even rogue Irish judges are aware of their public profile and have a limit as to how much in-your-face hubris and hypocrisy they will indulge in; and besides, as more-and-more others start using these same tactics, well, think of how judges will react when a member of the public asks them in open Court whether they have any criminal proceedings pending against them – and if so, well I'm very sorry judge, but we simply can't be dealing with criminals now, can we? Trapped by their own hubris and mendacities in other words.

Before anyone raises the parallel issue of enforcement and whose army exactly are we going to use to put all of these criminal miscreants where they truly belong; please remember that we too have to follow the Triadic Archetype formula of, (i) concept, (ii) plan, and (iii) action, so before we start worrying about training up the required 'Officers of the Tribunal' who will have the authority to serve legal papers and make citizens arrests on behalf of the PTI Council for example, let's just focus on getting the paperwork right first, and getting all of these legal ducks in a row, and upon that foundation we may then start working on the next steps. But still, we have already produced a handful of I-I Asseverations and PTI Determinations that anyone can use in their own circumstances. So, as soon as this book is published we will be devoting our energies into the production of more and more Covid-related Asseverations as blanket protections against the criminal intentions of the Cabal.

For example, and just like in many other countries one would imagine; the whole justification for the Covid emergency legislation in Ireland for the granting of extraordinary powers to the Minister for Health hinges on the following clause in the *2020 Covid-19 Heath Act*:

> *"Having regard to the immediate, exceptional and manifest risk posed to human life and public health by the spread of Covid-19.."*
> cxxxvi

Well, what if we could demonstrate, literally and lawfully – perhaps by using the contents of this book as an exhibit – duly backed up by respective Asseverations – that there is NO, "immediate, exceptional and manifest risk" here? Where would that leave Minister Donnelly and his diktats? It would pull the proverbial rug right out from under them and they – and all of their unlawful directives – would be left exactly where they belong; in the history books as part of a massive failed experiment in social engineering, to be used in evidence at the Nuremberg 2 Trials. And if anyone is in any doubt that this Covid legislation is totally without merit, legitimacy or legal validity, please just get yourself a copy of the INDICTMENT booklet which explains how the Irish Courts are pulling a massive legal scam on the people, and they have been doing this for decades in full knowledge amongst all of the members of the Irish establishment (because we keep on sending them the proofs) that the Irish Courts have no legal jurisdiction or authority to be conducting any sort of 'business' whatsoever. But then again, it's always been about the Great Lie.[15]

Hopefully, as everyone can see in the following short summary, a big part of this new peoples-justice approach is in publicly naming, shaming, exposing, confronting, denouncing and then 'proscribing' these individuals so that they finally begin to realise that there WILL be consequences in the court of public opinion at least, if they continue their nefarious behaviours, and that we are determined to make it as difficult and uncomfortable as we can, for these knaves and villains to continue their underhanded activities with impunity.

THE HAFTA PROCEDURE

1. Identify a single, definitive wrong that has been done by an authority figure and find and quote the corresponding law that has been violated.

2. Check to see if an I-I Asseveration already exists to cover that specific offence, and if not, then use the template to research your question and return your findings to fca@integrityireland.ie so that the I-I Commission can raise a new Asseveration for that

---

[15] INDICTMENT & Petition for a Public Enquiry Into State-Sponsored Criminality in Ireland.

question or offence – to help you, and others in future.

3. Complete a 'Statement of Truth & Fact' [or affidavit] (as short as possible) [see p.292, in the *I-I S.O.S. Guide*] and quote the law that has been broken and the stated penalties for doing so.

4. 'Price the crime' by putting a value on whatever loss has been caused to you, and formally bill/invoice the Accused; to be paid within a certain time frame. This opens the way for a claim for costs and damages and/or a Small Claims Court civil action.

5. Lodge your 'Statement of Truth' as a formal criminal complaint with An Garda Síochána, and collect a Pulse Number for that complaint.

6. Serve the accused with the NOTICE OF SUBMISSION OF A CRIMINAL COMPLAINT [p.281, I-I S.O.S. Guide] complete with copies of your Statement of Truth / the respective Asseverations / and any other related materials, either; (a) by 'Under Seal' (free) recorded post and/or; (b) by email, and/or; (c) by hand, duly witnessed.

7. Prepare a 'common informer' application using Form 15.3 Information and Form 15.1 Summons [p.278 & 279, I-I S.O.S. Guide] and lodge them with the District Court with your Statement of Truth / Asseverations etc. See *"D.I.Y. Justice in Ireland"* booklet for details. But remember to sign and date the foot (or back) of the Information Form on the day that you first send or take it into the District Court.

8. Serve a 'NOTICE OF PROSCRIPTION & EXCLUSION' on the accused and copy the same to their direct superior(s) / parent organisations / and to the I-I Association.

9. Alert the accused's public liability insurers and any other 'interested parties' that criminal proceedings have issued against them, copying the respective documents.

10. If the statutory authorities do NOT conduct investigations or issue the requested 'common informer' summonses within the

stated timeframe (which refusal would render any such offending person, in turn, subject to this HAFTA procedure) then place the accused formally 'On Notice' using the NOTIFICATION OF AN IMPENDING CITIZEN'S ARREST (if applicable under the 1997 Act) accompanied by a copy of the CONSTITUTIONAL NOTICE OF SERVICE ON ALL IRISH AUTHORITIES of December 4th 2020. [QTC No 6]. These 'QTC' Notices (silence implies consent) can be found in the *Criminality in the Irish Courts'* book.

11. Chose a time and place to perform the citizen's arrest and request a Garda escort to take custody of the accused.

12. If there is no response or the Gardaí do not provide the escort, then proceed to initiate a verbal citizen's arrest of the accused in person, and direct him/her to deliver themselves to the Garda Station immediately. If they fail or refuse to do so, then, using the minimum amount of physical force necessary, transport the accused to the Garda Station to be transferred into the custody of An Garda Síochána. All aspects of the procedure must be filmed for everyone's protection.

Additional options that can be deployed – but only AFTER the service of the Notice of Proscription & Exclusion.

Arrive unannounced at the home (or business) of the accused accompanied by a small detail of around 7-12 persons. Target their neighbours' houses with 'Public Notice' flyers [p.296, *I-I S.O.S. Guide*] before assembling on camera in front of the property complete with posters / banners / badges etc., to read out the flyer. The footage to be posted online later.

(Optional) Approach the property (on camera of course) and serve whatever documents are appropriate, asking for the accused to accompany the I-I Detail to the Garda Station immediately.

Please note that these 'operations' should be kept confidential and confined to small groups of trained-and-trusted 'deputised' individuals who sign a disclaimer before participating. The operation should be completed as quickly as possible, so as to avoid any possible confrontation with Gardaí (who may have been alerted to an alleged "disturbance"). Those on the Detail should remain courteous, civil and dignified throughout.. ...and so on.

In time, if enough of us apply these tactics with courage and conviction in alignment with our own personal values and principles that we absolutely *refuse* to abandon, then it is only a matter of time before these corrupt kingdoms, cabals and hierarchies *must* fall. We know this, because we have already done it in limited ways in our own sphere. As we write this book for example there are the 100 officials and office holders publicly named in the *'Criminality In the Irish Courts'* book who dare not try to engage with us in any official capacity, because as soon as they try to do so, we return to their superiors asking one simple question: "Are you directing me (Judge/Minister/Taoiseach) to engage with this person in spite of the crimes he/she has been committing?" ...and then we send copies of whatever proofs or complaint statements to that superior, including a digital copy of the Report that has another 99 high-profile names in it, reminding that superior explicitly; that now that he/she has personal knowledge of the subordinate's crimes – as well as all of the other 99 – that according to the reporting obligations of *the Criminal Justice Act 2011* (or whatever similar legislation exists in your country) the superior is now <u>legally obliged</u> to report the crimes to the statutory authorities. And when they don't, well, now we have another name for the HAFTA procedure, and another name to go on the 'Proscribed & Excluded' list... and so on! You see, it's only a matter of time.

As I sit here writing this book for example, over a third of the sitting judges in Ireland cannot (lawfully) and WILL not deal with us, because to put it colloquially, we have too much dirt on them and their colleagues. Dirt that must inevitably be brought into evidence if they try any more of their dirty tricks on us. Amongst those judges are the current Chief Justice Frank Clarke and the man tipped to take his place later this year; Donal O'Donnell. Hopefully by the time Frank retires with his gilded pension and Donal slides in as new ringmaster people will have started asking, "But what crimes have Frank Clarke and Donal O'Donnell been accused of?" (and) "So... what happened to those complaints to Gardaí and the since-disappeared applications before the Courts?" (and) "Why hasn't the Justice Committee and the Minister properly dealt with the allegations and proofs that the holders of the highest judicial office in the land are in fact criminal miscreants with no respect for the law or the Constitution?" (and) "What happened to that new Judicial Council set up in July 2020 after 25 years of inexplicable delays. You know, that eminent body of arbitrators who are supposed to look into allegations of judicial misconduct? The same Council that has a majority of judges on it, including the two just named?" (and) "Why have the Clerks of the Dáil and the Courts Service embarked upon a collusive trail of lies, deceptions, stonewalling and disappeared documents..." – with the active complicity of the MSM in the form of *The Irish Times*, *The Western People* and other compromised news outlets to try to cover all of this up? And what are they going to do about this now eh? Or more specifically, what are they LAWFULLY going to do about it?

The answer is nothing! Absolutely nothing! They have NO lawful response to this type of straight-talking, in-your-face refusals to pander to career criminals and it's a tactic that any of us can use. What we're basically saying to them is, please give me the service and the respect that is due to me under the law, and if you can't, won't or don't do so, well, then you're not really a judge then, are you? And I have an Asseveration right here judge (or garda/civil servant/TD) that says that there is no obligation on me to knowingly engage with criminals. Indeed, it would be a crime for me to do so, would it not? So I'm really sorry, but regardless of who you think you are, there isn't a soul on this planet who has the right and authority to order me to commit a crime. So, you may now step down judge while we prepare criminal charges against you, and please, go and find us an honest replacement (if that's not too much to ask) and we will take it from there. Can't find one judge? Oh dear. We thought that might be the case. Ah well, just as well that the Irish People have set up their own Tribunal, because clearly, "the *ordinary courts are inadequate to secure the effective administration of justice.*"

As more and more people realise the futility and downright stupidity of sitting next to known pickpockets on the train, or of returning to dodgy mechanics time-after-time, or of paying for a second helping of poisoned soup; and are truly getting the simple fact that this is literally, mad, insane, psychotic and arguably criminal behaviour on our part; because even as their would-be victims, we are *still* actively facilitating and *knowingly* participating in their crimes are we not?

This is Stockholm Syndrome gone batty, stupid, gullible and daft all in one go! But on the other hand if we simply say 'NO'! If we tell them bluntly that I for one will NOT join them in their personal little Hell, and that they have NO power to order me to do so; not legally, not morally and certainly not without my full consent! Then everyone will start to realise that the sooner we stop returning to these criminals – to these so-called 'authority figures' – and the sooner we stop asking them for anything and start *telling* them instead; then the sooner the liars, the deceivers, the despicables and the diabolicals become irrelevant and redundant! We cut them out of the equation altogether in other words, unless they are willing to deal with us honestly according to the law. Because now we have alternatives. There *are* safe seats on the train now. We have also sourced properly-trained, honest mechanics, and a new restaurant is opening up as well, in the form of *The Peoples Tribunal of Ireland* and whatever other similar initiatives that spring up around the world. By taking firm and consistent actions like this, we will make the diabolicals irrelevant and redundant, by simply refusing to dance with the Devil whenever the tune does not resonate with our own core values and principles; such as honesty, truth and justice that align with social ethics and morals, which in turn reflect Universal Law.

CHAPTER TWENTY-SIX

A NEW BEGINNING?

Now all of this sounds great on paper, I know, and it will be a different matter altogether reimagining a new type of society that is modelled on the themes of the Triadic Archetype. But not surprisingly, this is exactly how we have to start. Because we must first of all get the *concept* right before moving to the plan and then to action. And the concept in this case is in first of all understanding that the original model for human society that was conceived in the Garden of Eden has been utterly corrupted and malformed by the predatory oppressive masculine in history with all of his diabolical structures and hypocritical hierarchies, and whatever we imagine or rebuild from here, it simply cannot be a replication or duplication of the same old 'Satanic' model which suppresses the feminine characteristics of maternal wisdom; that rewards ruthless, inhumane profiteering; and is almost completely devoid of the Spirit of Truth.

Of most pressing concern perhaps is the need to ensure that no more soul-less sociopaths and psychopaths are gifted positions of power and authority in traditional socio-political hierarchies, which means that we are going to have to reimagine and rethink what form future societies should take. This will take time of course, but the very first step in that process is training ourselves to spot the diabolicals and simply refuse to have dealings with them. Don't encourage them, don't work for them, don't fund them and most certainly don't ever vote for them again.

In respect of a balanced, properly-functioning future society, the Triadic Archetype model would be based upon the principle that the individual finds purpose and meaning by serving and contributing to the collective, which in turn operates according to Universal Law. Or conversely, that Universal Law informs the social-collective how best to nurture and protect the individual membership of the community. In such societies, well-informed empaths would naturally rise to positions of truth and respect based upon their proven worth to the community – again – not necessarily in material, mechanical or profit-related terms, but based upon their knowledge, understanding and alignment with Universal Laws, and on their service to the collective.

Concept	Universal Law		Ideology	Spirit
Plan	The Collective		Wisdom	Mind
Action	Individuals		Daily Activity	Body

Such future societies would draw upon the wisdom of proven 'Elders' who would take their natural place in communities that respected wisdom, compassion, truth and proven life experience over fast-talking whippersnappers in sharp suits who think they know it all, even before they've had the chance to learn a few crucial life-lessons such as the value of serving the collective and the necessity to put the common good before selfish aims and ambitions. Truth would be placed at a premium, and would be recognised as what it is; as THE most valuable commodity of all, and young people would be encouraged to explore, and test, and hypothesise in that direction.

In contrast to today's world where the most efficient liars and deceivers get rewarded with money, power, position and prestige, those who would engage in deceits or deceptions will be recognised and effectively marginalised in society until they reformed their ways. There would be no rewards in other words, for the usual sociopathic traits and behaviours, for lies and false narratives, and anyone who attempted to do so would simply find themselves being abandoned by the decent bulk of society, with people simply saying 'NO' – we will not partake in your sins and deceptions – not as an employee, not as a partner, not as a consumer, and not even as passive bystanders any more. We simply will not allow it! The liars must from now on be left to their own.

343

Political parties and institutions as we have known them, with their intrinsic deceptions and self-interests will – and indeed must – become a thing of the past. These self-centred political cliques will be replaced by small, short-term cyclical rotations perhaps of ordinary members of the community who will serve in local councils for a predetermined period, with regional committees comprising persons from those local councils, with a national assembly of representatives from those regional committees perhaps? Not elected mind you, but chosen at random in rotation so that everyone will get a chance during their lifetimes to genuinely serve their own communities at some level whilst contributing their personal expertise to the collective. At the same time we will all get to know diverse others and gain understandings in the inner logistics of getting collective projects done efficiently and creatively – but always with an eye on what gives the community true value. Not money, not things, not positions of leadership and power, but simply and essentially, the enhanced quality of all our lives that will make people naturally happy and therefore more naturally inclined to want to participate, and to share.

'Wages' for these periods of public service – if we really need to go that route – would be modest and linked to the economy and to the average working wage, so that anyone operating as a genuine public servant, could not – and would not – be distracted by temptations to abuse one's role. In fact, if we get the selection formula right, the type of people serving as councillors or assembly members would not be inclined to systemic dishonesties or exploitations anyway.

Perhaps we should introduce a new set of qualifications and criteria for certain important roles and positions in society, with the preeminent qualifier being that of morality and decency. In such a world, the sociopaths and psychopaths would be filtered out early in life and excused any role or participation in any form of governance. That is not to diminish the constructive roles that clinical psychopaths could still perform including high tech, mechanical or mathematical jobs that require some of the characteristics of surgeons for example who are placed 4th on the psychopathic list of occupations after CEO's, lawyers, and media personnel, because surgeons are better off not crying into open wounds or getting all emotional during life-saving surgeries and need to maintain a mechanical objectivity about the task before them. We just wouldn't place a surgeon in charge of hospital admissions policy, because even if he is a decent, well-functioning member of the community who is excellent at his job, he will not (if he is a sociopath or a psychopath) have the empathetic attributes to make humane, vs purely mechanical-biological-fiscal decisions. Accountants, IT experts, scientists, clowns and general labourers are also the type of jobs that sociopaths could be doing productively without being a danger to themselves or to others, and, if we get the formula right, those who have not been permanently and irreversibly conditioned into psychopathy may eventually re-grow a conscience and a spirit of community sufficient to broaden their horizons and move to more integrated occupations and careers.

Elders meanwhile, could assume all of the major governance roles as human societies transitioned from hierarchical tyrannies where 'the boss' dictates what's to be done, to socially-inclusive communities that are based upon collective, practical wisdom in alignment with Universal Law. On the assumption that we will still need them – at least for a while – or at least until initiatives like *The Peoples Tribunal* and such like reinvent a workable version of justice; then judges too would have to demonstrate a genuine capacity for wisdom and compassion before being allowed anywhere near a Courtroom. Lawyers and other such 'officers of the Court' could still be sociopaths and psychopaths of course (because we're going to have to put them somewhere) as long as the presiding judge is not. Because we will definitely need firm, wise and empathetic figures overseeing traditional Court proceedings if we are to ensure that justice will be properly done. In fact, it would be a truly novel experience would it not, to see a little justice being done in the Irish Courts even if only as part of this evolutionary transition? Because arguably, the need for courts of law and such like will also diminish as we each take responsible ownership for our actions, and where the usual rewards for criminality will seem like cheap change in comparison to living whole and fruitful, honest and truly happy lives.

Private ownership of 'stuff' would become less and less important as more of us cooperated in sharing, lending and borrowing whatever we needed in a spirit of collective understanding that possessions in-and-of themselves only have value in the way we use or deploy them in good service to others. Those fortunate enough to own land, or property, or machinery, or vehicles, or businesses and the like, would be rethinking how to deploy those assets in a manner that returned maximum benefit to the collective whilst maintaining the quality of those assets. Big business – in the forms we have it today with the bosses of international conglomerates being of necessity, profit-driven sociopaths and psychopaths – would gradually lose their appeal, with enlightened workers seeking employment in more ethical operations with less of a monopoly and more of a local-service ethos that reciprocates organically with the community.

The capitalist mindset too will eventually be abandoned as more and more investors accept and understand that they are profiting personally, but also amorally, on inhumane investments that come at the expense of others who are the victims of the profiteers.

Foods would be grown and harvested with a focus on nutrition – not profit – and nature would be respected again as our earthly mother and as the source of all things essential. Produce would be shared and consumed locally with any excess being made available to the local Council for anyone in need – and without charge. All adults would commit to giving of their time and particular expertise in some sort of service to the community – even if such had no intrinsic productivity-related 'monetary value' such as music, storytelling or the arts for instance, with equal portions of time being spent in labour and in recreation. Associations and collectives could be formed for the advancement of the arts and the sciences for example, as long as those organisations reported to their regional or national assemblies who would in turn ensure that those projects remained 'on track' for the betterment of society in alignment with humane principles. All of the hitherto great advances in science in particular – and all of the apparatus and machinery that supports scientific endeavour – could thus be deployed, NOT in the exclusive pursuit of profit as is currently the case, but in the great and everlasting quest for knowledge and understanding of the world around us only this time, again, in alignment with Universal law and in context of humane principles.

Freedom would be an elevated principle. Freedom of thought and expression; freedom of movement and travel; freedom to choose our own vocations and occupations as long as they met some need of the collective. The freedom even to violate Universal Law would remain – just as that same freedom of choice existed in Eden – but those that chose to do so would this time understand that any such actions would only alienate them from the community. Whatever local 'laws' may be necessary would be policed and enforced with the minimum consequences to maintain peace, and always with an eye on correction-and-reform vs penalty or punishment.

All things natural – including the healing arts and medicine – would be relearned and re-taught from a perspective where we understand our roles as the stewards, and not the owners of the Earth, and where we reimagine the integrated place of that historically-interlinked organism that we call 'humanity' inside a billion-year-old ecology that continues to evolve and adapt despite our best attempts so far, to destroy it.

Education would no longer be the systemic conditioning and indoctrination of the masses for the purposes of servicing the globo-corporate-political Machine, but would be formed and shaped in such a manner as to give genuine liberty to young and enquiring minds to explore the world around them in all of its beautiful and complex manifestations. Practical education would include children learning all of the necessary life skills to be able to survive and thrive, and to participate fruitfully in the community through apprenticeships or other guided activities by skilled mentors and wise Elders. Travel to other countries, cultures and locations would be seen as essential to genuine education and would be interwoven into our children's schooling – whether institutionalised or not – with all communities acting as foster-hosts to foreign children and even whole families to improve global cooperations and understandings.

Money – oh I don't know – I wish we could get rid of it altogether and perhaps replace it with a merit-credit system whereby all new children being born would be 'credited' with a universal amount of life-tokens or whatever we might call them. These life-tokens would be sufficient to ensure that everyone had the essentials in life – for the whole of their lives – including access to food, shelter, education, medical care etc., etc., and one could enhance one's own credit standing by doing good works in the community. It would only be people who had 'earned' the right through decent, unselfish behaviours who would be eligible for selection to local Councils or national assemblies or to be honoured as Elders. This would NOT of course be a punitive merit system such as exists in China today, where citizens can effectively be punished by an all-surveilling State for any perceived transgressions, but would be solely based upon rewards for constructive, productive behaviours.

Provided we had the right type of wise and empathetic people (Elders?) maintaining overall social cohesion based upon a genuine collective spirituality rooted in kindness, compassion, truth, respect and mutual collaborations; then there is no reason to doubt that we could recreate human society in a manner that would be a perpetual cycle of creativity, of cooperation, of innovation, of mutual respect, of affection and love, and indeed, of many of the concepts contained in the ideas of a Great Reset and of a New World Order, except only this time, for all of the *right* reasons.

*   *   *

These are just a few ideas of how we might rethink the beginnings of a new state of existence for humanity whereby we need not be wracked by guilt or shame about our history thus far nor worried about the diminishing prospects for our children; nor about the behaviours and intentions of billionaire psychopathic visionaries; but where we can be fully and wholly invested in creating a world that will be a joy to live in with full expression of the spirit, the mind and the body in all of the delightful and wonderful ways that we humans were originally designed to express.

Evil, as an interference in Universal, Natural Law would have no place in such societies. There would be plenty of room of course for all manner of differences and approaches and outcomes – some better than others depending upon who was doing what, and how well they were doing it – but all of these myriad efforts would be centred upon a spirit of truth and of collective good, so, only good could come of it. With all people focusing upon what is right and good, the only differences in outcomes would be due to our various individual perceptions of how best to embody truth and goodness in our lives, and we would be constantly learning from each other in so many diverse, but positive ways.

In short; evil has to go. And we – the people who are here, alive today – are the only ones who can do it. This global crisis is the opportunity either for evil to finally succeed with all of the now-predictable horrors that will result, or, for good to triumph over evil through the agency of collective courage and resolve.

By dealing effectively with the phenomenon of human evil we will create the space for the Divine to re-enter human societies. The concept is already there – as provided by Universal Law. It is now up to the enlightened empaths, teachers and thinkers out there, to make a wise and workable plan, and then for the rest of us – the good and the noble and the dedicated amongst us – to simply make it happen!

ORIGINAL LOVE

The good Lord called in a voice full of joy...
"HALLELUJIAH! – A new creation!
Come hither ye three, and decide who should be
Protector, Director and Teacher."

"This Adam you see.., will be head of the tree
Of the family who knows love and life,
And I need you to ensure that his soul remains pure
While I go and create him a wife."

Now the angels could see that the Lord was real pleased
With his 'image and likeness' incarnate
And as they chose who should be, guardians for Adam and Eve
Poor Lucifer became quite distracted..

For as "Knight of the Light" he was exceptionally bright
And quickly deduced the equation;
He has made these to be in dominion over me
And all of the rest of creation!

I just can't believe He would do this to me,
I thought I was His dream of perfection.
But never for me has He created an Eve
Why should Adam get all this attention?

So, feeling dejected and somewhat rejected
He dismissed poor old Michael and Gabe
Took the job of Protector, Director and Teacher
And broodily waited for Eve.

Now, Eve was a beauty – all naked and lovely
...and curious, and rather naïve
Saw her angel-protector, magnificent in splendour
And surrendered to his sweet deceit.

And on his insistence she took that sweet fruit,
The produce of that dark betrayal
And shared it with him who was born without sin
That first son of our God and Creator

And so was begotten the first son of Satan
The first one to turn against God.
Their arrogant treason poisoned sweet Eden
And corrupted original love.

Books like this are being suppressed by the establishment. So if you have enjoyed reading this book – or appreciate its contents – we would be most grateful if you could spread the word with a customer review on Amazon, the Great British Bookshop or elsewhere.

Other books worth reading covering the themes in this book include:

- "CORONA FALSE ALARM? Facts & Figures" by Dr Karina Reiss & Dr Sucharit Bhakdi.
- "How The World Works" by Naom Chomsky.
- "The Impact of Science on Society" by Bertrand Russell.
- "Anyone Who Tells You Vaccines Are Safe & Effective is Lying" by Dr Vernon Coleman.
- "The Coming Apocalypse" by Dr Vernon Coleman.
- "HUMANTRUTH. A Philosophy For A World In Crisis" by John Bapty Oates.
- "NEW WORLD DISORDER. The Case for Economic Democracy, Reformed Political Democracy & A Coming Together of All People" by David Egan.
- "CORRUPTION. What Everyone Needs To Know" by Ray Fishman and Miriam A. Golden.
- "Global Discontents" by Naom Chomsky.
- "Your Celtic Path" by Con Connor.
- "THE ESTABLISHMENT. And How They Get Away With It" by Owen Jones.
- "THE RULE OF LAW" by Tom Bingham.
- "My Awakening and the Covid-19 Fraud" by Ciaran Boyle.
- "Exposition of the Divine Principle" by Dr. Won Pok Choi.
- "COVID: What Public Health Won't Tell You" by Carina Harkin.
- "The Colour of Truth Vol I: Amazing Coincidence? ...or Intelligent Design?" by Stephen T Manning.

References: For this list in digital form, please visit the 'books' tab on the Integrity Ireland website.

[i] See, "A Very British Conspiracy" by John Dekker. ISBN 9781906628543.

[ii] Criminal Litigation, 3rd Ed. pp 15-20 'Regulatory Crime'. The Law Society of Ireland.

[iii] List of Murdoch's News Corporation's Major Holdings | Voice of America - English

[iv] https://schoolhistory.co.uk/notes/operation-mockingbird/

[v] https://text.npr.org/812499752

[vi] https://en.wikipedia.org/wiki/Psychopathy_in_the_workplace

[vii] https://www.mic.com/articles/44423/10-professions-that-attract-the-most-sociopaths

[viii] https://en.wikipedia.org/wiki/Timeline_of_the_COVID-19_pandemic_in_the_United_Kingdom_(January%E2%80%93June_2020)

[ix] Were Dead Bodies Left On Wuhan Streets After The Coronavirus Outbreak? (boomlive.in)

[x] Perma | birdofprey on Twitter: "The fallen put outside ..we need the truth! #truth #CoronavirusOutbreak #corruption... "

[xi] https://www.ndtv.com/india-news/videos-of-dead-bodies-were-fake-indian-student-who-returned-from-wuhan-2190884

[xii] https://www.jstor.org/stable/40225240?seq=1

[xiii] U.S. National Debt Clock : Real Time (usdebtclock.org)

[xiv] https://www.bonkers.ie/blog/banking/negative-interest-rates-here-s-everything-you-need-to-know/

[xv] The Fauci COVID-19 Dossier | Principia Scientific Intl. (principia-scientific.com)

[xvi] Dr Mike Yeadon's Final Warning to Humanity – Rights and Freedoms (wordpress.com)

[xvii] https://www.bbc.com/future/article/20200617-what-if-all-viruses-disappeared

[xviii] https://www.medicalnewstoday.com/articles/158179

[xix] https://www.who.int/health-topics/severe-acute-respiratory-syndrome#tab=tab_1

[xx] https://www.cidrap.umn.edu/news-perspective/2003/05/estimates-sars-death-rates-revised-upward

[xxi] https://www.cdc.gov/flu/pandemic-resources/1918-commemoration/pdfs/1918-pandemic-webinar.pdf

[xxii] https://foreignpolicy.com/2019/09/20/the-world-knows-an-apocalyptic-pandemic-is-coming/

[xxiii] https://foreignpolicy.com/2019/09/20/the-world-knows-an-apocalyptic-pandemic-is-coming/

[xxiv] https://www.youtube.com/watch?v=GWPIhgknRhl

[xxv] The Definition of "Pandemic" has been Altered – Undercurrents (wordpress.com)

[xxvi] https://www.rte.ie/news/health/2020/0703/1151127-virus-report/

xxvii[xxvii] https://twitter.com/leovaradkar/status/1278995351169613824?lang=en

xxviii https://en.wikipedia.org/wiki/Koch%27s_postulates

xxix Mirrored | Dr Andrew Kaufman | Koch's Postulates... Have They Been Proven For Viruses? - Bing video

xxx https://www.nejm.org/doi/full/10.1056/nejmoa2001017

xxxi See p.39 in the CDC Report entitled: "CDC 2019-Novel Coronavirus (2019-nCoV) Real-Time RT-PCR Diagnostic Panel."

xxxii https://www.eurosurveillance.org/content/10.2807/1560-7917.ES.2020.25.32...

xxxiii Laboratories in US can't find Covid-19 in one of 1,500 positive tests (greatreject.org)

xxxiv PLANDEMIC – INDOCTORNATION - The Documentary - YouTube

xxxv The Fauci COVID 19 Dossier : David Martin : Free Download, Borrow, and Streaming : Internet Archive

xxxvi Helping People Be Seen, Heard and Believed After Adverse Vaccine Reactions - Ron... (senate.gov)

xxxvii https://www.facebook.com/watch/?v=1131775030302811

xxxviii https://www.youtube.com/watch?v=IGteqzE6qmM

xxxix Global Preparedness Monitoring Board (who.int)

xl https://www.centerforhealthsecurity.org/our-work/events/2018_clade_x_exercise/

xli Rockefeller-Foundation-2010-Scenarios-for-the-Future-of-Technology-and-International-Development.pdf (truthcomestolight.com)

xlii Bill Gates killed poor tribal Indian girls through PATH "vaccine" initiative | The Planet Today News From The World. (planet-today.com)

xliii http://thepeoplestribunalofireland.ie/index.html

xliv https://indianapublicmedia.org/eartheats/bill-gates-monsanto-team-world-hunger.php

xlv https://dta0yqvfnusiq.cloudfront.net/allnaturalhealingsrq/2019/04/How-Rockefeller-Founded-Big-Pharma-and-Waged-War-on-Natural-Cures-5cb3d7374f337.pdf

xlvi https://www.independent.ie/entertainment/books/book-reviews/new-jeffrey-epstein-book-reveals-the-medias-conspiracy-of-silence-protecting-hissexual-pyramid-scheme-40726592.html

xlvii https://www.youtube.com/watch?v=cmbRR-JGHnQ

xlviii https://fullfact.org/about/our-team/

xlix https://www.mayoclinic.org/diseases-conditions/coronavirus/in-depth/coronavirus-myths/art-20485720#

l https://fullfact.org/about/our-team/

li https://www.thejakartapost.com/news/2020/10/04/pcr-tests-remain-gold-standard-but-many-factors-affect-results.html

lii https://www.ncbi.nlm.nih.gov/pmc/articles/PMC7204879/

[liii] https://www.thelancet.com/journals/lanres/article/PIIS2213-2600%2820%2930453-7/fulltext

[liv] https://bpa-pathology.com/covid19-pcr-tests-are-scientifically-meaningless/

[lv] Ethylene Oxide - Cancer-Causing Substances - National Cancer Institute

[lvi] https://www.ncbi.nlm.nih.gov/pmc/articles/PMC7961671/

[lvii] https://www.tribpub.com/gdpr/nydailynews.com/

[lviii] NEED TO WATCH. The Dangers of Ethylene Oxide in the PCR Test - Bing video

[lix] https://www.who.int/publications/i/item/10665-331501

[lx] Newly-released Emails Reveal Dr Fauci Spoke With Bill Gates Regularly, Knew Masks Don't Work, and Was Aware of Lab-Leak Likelihood – DailyVeracity

[lxi] https://vimeo.com/437157465

[lxii] How Big is a Virus? | Exploratorium - YouTube

[lxiii] How Far Do Sneezes and Vomit Travel? - YouTube

[lxiv] WHO Expert Member Peter Daszak Seen Boasting Of Manipulating Killer Virus In China In 2016 - Bing video

[lxv] https://www.youtube.com/watch?v=ZfOjvfjFwOg

[lxvi] List of COVID-19 vaccine authorizations - Wikipedia

[lxvii] https://fullfact.org/health/Covid-QA-graphic/

[lxviii] Pfizer-BioNTech COVID-19 Vaccine EUA Fact Sheet for Recipients and Caregivers (fda.gov)

[lxix] Janssen COVID-19 Vaccine EUA Fact Sheet for Recipients and Caregivers 07122021 (fda.gov)

[lxx] https://journals.plos.org/plosone/article/figure?id=10.1371/journal.pone.0032857.g004

[lxxi] https://www.cdc.gov/vaccines/pubs/pinkbook/downloads/appendices/b/excipient-table-2.pdf

[lxxii] Poisonous Ingredients In Vaccines - VaccineFromHell.comVaccineFromHell.com

[lxxiii] Helping People Be Seen, Heard and Believed After Adverse Vaccine Reactions - Ron... (senate.gov)

[lxxiv] FRIGHTENING! – 16th update on Adverse Reactions to Covid Vaccines released by UK Government / MHRA – Rights and Freedoms (wordpress.com)

[lxxv] https://www.naturalhealth365.com/no-covid-jab-mandate-for-pfizer-3947.html

[lxxvi] WO2020060606 CRYPTOCURRENCY SYSTEM USING BODY ACTIVITY DATA (wipo.int)

[lxxvii] The Jansen vaccine was removed from the Irish market on the 4th of August (rumble.com)

[lxxviii] https://en.wikipedia.org/wiki/Human_coronavirus_229E#cite_note-20

[lxxix] https://www.forbes.com/sites/alexknapp/2020/04/11/the-secret-history-of-the-first-coronavirus-229e/

[lxxx] https://www.fda.gov/consumers/consumer-updates/why-you-should-not-use-ivermectin-treat-or-prevent-covid-19

[lxxxi] https://www.thelancet.com/journals/laninf/article/PIIS1473-3099(11)70065-2/fulltext

[lxxxii] https://www.news-medical.net/news/20200406/Antiparasitic-drug-Ivermectin-kills-coronavirus-in-48-hours.aspx

[lxxxiii] https://www.brisbanetimes.com.au/national/queensland/doctors-banned-from-prescribing-potential-covid-19-drug-20200408-p54ic1.html

[lxxxiv] https://www.who.int/news-room/feature-stories/detail/who-advises-that-ivermectin-only-be-used-to-treat-covid-19-within-clinical-trials

[lxxxv] https://www.bbc.com/news/51980731

[lxxxvi] Ibid.

[lxxxvii] https://www.contagionlive.com/view/zoonotic-threats-as-unpredictable-as-they-are-dangerous

[lxxxviii] https://www.contagionlive.com/view/zoonotic-threats-as-unpredictable-as-they-are-dangerous

[lxxxix] https://www.youtube.com/watch?v=0IV-FXRttdl

[xc] https://en.wikipedia.org/wiki/Antibody-dependent_enhancement

[xci] https://trialsitenews.com/indian-protein-based-vaccine-for-covid-19-enters-phase-3-developed-by-baylor-college-of-medicine/

[xcii] https://www.washingtonpost.com/wp-srv/style/longterm/books/chap1/inventingtheaidsvirus.htm

[xciii] https://patents.google.com/patent/US6372224B1/en?oq=6372224

[xciv] https://patents.google.com/patent/US6593111B2/en?oq=+6%2c593%2c111

[xcv] https://patents.google.com/patent/US7279327B2/en?oq=7279327

[xcvi] https://patents.google.com/patent/US7220852B1/en?oq=7220852

[xcvii] https://patents.google.com/patent/US7776521B1/en?oq=7776521

[xcviii] https://patents.google.com/patent/US7151163B2/en?oq=US-7151163-B2

[xcix] https://patents.google.com/patent/US7618802B2/en?oq=7618802

[c] https://www.m-cam.com/wp-content/uploads/2020/04/20200403_SARS_CoV_Patent_Corpus_Lit_Review.pdf

[ci] https://www.biorxiv.org/content/10.1101/2020.03.02.974139v2

[cii] https://www.nature.com/articles/s41591-020-0820-9

[ciii] https://www.wsj.com/articles/SB105209016979730900

[civ] Reverse genetics with a full-length infectious cDNA of severe acute respiratory syndrome coronavirus | PNAS

[cv] https://www.chamberlitigation.com/sites/default/files/David%20W%20Rowell%20v%20Montana%20Governor%20Bullock.pdf

[cvi] https://www.jfklibrary.org/asset-viewer/archives/JPWPP

[cvii] http://www.integrityireland.ie/page61.html

[cviii] http://thepeoplestribunalofireland.ie/index.html

[cix] https://www.weforum.org/

[cx] https://www.weforum.org/focus/the-great-reset

[cxi] https://www.cnbc.com/2019/11/20/bulletproof-coffee-founder-dave-asprey-how-to-live-longer.html

[cxii] Nursing home deaths 2021 - Health Freedom Ireland

[cxiii] Letter sent to government re dramatic rise in excess nursing home deaths - Health Freedom Ireland

[cxiv] The Year of Merciless Killing - by John Waters - John Waters Unchained (substack.com)

[cxv] Covid Vaccines: 'One small step for the virus, one giant catastrophe for mankind' - by John Waters - John Waters Unchained (substack.com)

[cxvi] https://undercurrents723949620.wordpress.com/2021/05/22/how-many-have-died-from-covid-vaccines/

[cxvii] https://undercurrents723949620.wordpress.com/home/

[cxviii] http://indymedia.ie/vaccine%20deaths

[cxix] https://www.cacatholic.org/CCC-vaccine-moral-acceptability

[cxx] https://www.abortionrightscampaign.ie/2020/09/23/what-happened-during-irelands-first-year-of-legal-abortion/

[cxxi] https://www2.hse.ie/wellbeing/mental-health/supports-for-young-people.html

[cxxii] https://www.congress.gov/bill/116th-congress/house-bill/6666/titles

[cxxiii] https://patentscope.wipo.int/search/en/detail.jsf?docId=WO2020060606

[cxxiv] https://bpsbioscience.com/spike-sars-cov-2-pseudotyped-lentivirus-luc-reporter-79942

[cxxv] https://www2.hse.ie/screening-and-vaccinations/covid-19-vaccine/get-the-vaccine/when-you-have-been-vaccinated/

[cxxvi] Overview (gpmb.org)

[cxxvii] Freemasonry Proven To Worship Lucifer (bibliotecapleyades.net)

[cxxviii] http://holyspiritvictorious4ever.blogspot.com/2009/12/united-nations-plaza-located-at-666th.html

[cxxix] https://www.biospace.com/article/comparison-2003-sars-pandemic-vs-2020-covid-19-pandemic/

[cxxx] https://vaxopedia.org/2021/05/02/were-african-leaders-murdered-for-opposing-the-covid-19-vaccine/

[cxxxi] https://www.independent.ie/irish-news/health/6666-abortions-took-place-in-first-full-year-since-law-changed-39330420.html

[cxxxii] http://www.integrityireland.ie/page61.html

[cxxxiii] https://www.amazon.co.uk/INDICTMENT-APPLICATION-ENQUIRY-STATE-SPONSORED-CRIMINALITY/product-

[cxxxiv] https://www.amazon.co.uk/CRIMINALITY-IRISH-COURTS-Absence-Ireland/

[cxxxv] https://www.amazon.co.uk/D-I-Y-JUSTICE-IRELAND-Prosecuting-Informer/

[cxxxv] http://www.irishstatutebook.ie/eli/2020/si/352/made/en/print

We are all being sleepwalked into Hell. *THEY* want total control of the world and all its 'resources' – including you! The Covid-19 Agenda requires all of us to be injected with 'stuff' that will either render us controllable and manageable for the Machine or, that will kill us!

The authorities are NOT to be trusted. Make no mistake. This is the great and final battle between good and evil. The apocalyptic Armageddon is at hand! Each of us has to decide: do we comply with the worst crime vs humanity in history – or not!? Do we sell our souls or liberate our minds? The choice is ours. **Please, please…, choose wisely.**

Bulk orders of this book in multiples of 10 can be ordered directly, at a discount from bookstore@checkpointpress.com.

Please spread the word.

Lightning Source UK Ltd.
Milton Keynes UK
UKHW021156150222
398726UK00007B/1678